Suppliers to the Confederacy II
S. Isaac Campbell & Co., London
Peter Tait & Co., Limerick

By Craig L. Barry & David C. Burt

The Stainless Banner Publishing Company

www.thestainlessbanner.com

Craig wishes to dedicate this book to the "War Department."

David wishes to dedicate this book to the "memory of John Hopper, who inspired me to learn all I could about S. Isaac Campbell & Co."

CONTENTS

FOREWORD

Confederate agents in Europe faced an overwhelming task. Purchase and ship everything the armies would need to stay in the field. Large orders for guns, cannon, ammunition, clothing, shoes, swords and medical supplies were placed with London commission houses for immediate shipment. And, oh, by the way, there was not enough cash on hand to pay for the items being purchased. Yes, it was an overwhelming task.

Craig and David faced a similar task in writing the histories of the suppliers to the Confederacy. They had to hunt down sources spread out over two continents, ransack the Original Records for every correspondence between Great Britain and Richmond and discover treasures hidden away in unknown archives. Then they had to organize those sources into a coherent story, filling in the blanks when the primary sources were silent on what happened next.

It might have been an overwhelming task, but Craig and David have brilliantly succeeded. The stories of S. Isaac Campbell & Co. and Peter Tait & Co. are compelling to read and shed a hot, white light on the struggles the War Department faced in making sure that the armies were armed and dressed. Books upon books have been written about the Army of Northern Virginia and Army of Tennessee, but very few books written on the men who made it possible for Generals Lee, Bragg, Hood and Johnston to remain in the field, fighting against impossible odds, for four years.

Suppliers to the Confederacy II would be an important work in the war's historiography if Craig and David had just stuck to telling the story of the procurement of supplies. But, instead, they chose to use their experience and expertise and examine the items that were supplied. Buttons, swords, shoes, greatcoats are all highlighted. The book rushes to its crowning success – an in-depth analysis of the surviving Tait jackets. By time the authors finish their analysis, even I would be able to pick out a true Tait jacket from the myriad of imposters.

So, yes, Craig and David have brilliantly succeeded with their latest manuscript.

C.L. Gray
Editor, *The Stainless Banner E-Zine*

PREFACE

I first met John Hopper at a Civil War re-enactment in 1996. He was a talented historian and scholar who told us a story around the campfire of how the British had enormously aided the Confederacy with arms and equipment, and how S. Isaac Campbell & Co. had become the primary mover and shaker when it came to supplying the South with all its needs. Fascinated by the story, I chatted with John over the weekend. In the aftermath of many telephone calls and emails, I decided to find out more about S. Isaac Campbell & Co.

John had become obsessed with Samuel Isaac after working in Isaac's former factory in Northampton. Being an avid researcher, he wanted to know more about both Samuel and Saul Isaac and how they had aided the Confederacy during the war. Over the course of many months, we began to piece together the story with a view to writing a book on SIC & Co.

We delved into the Northampton local newspaper archives and libraries, and the Official Records of the Union and Confederate Armies, which had just become available online. The story was just coming into being when I learned that John had died. I felt I had to carry on for John's sake and finish the book.

In 2007, I was fortunate enough to visit the newly discovered McRae Papers in the South Carolina Confederate Museum & Relic Room in Columbia, South Carolina. Along with curator Krissy Dunn, I researched the extensive records collected by Colin McRae during his investigation of SIC & Co. Satisfied that I had enough material, I self-published my first attempt on the story of SIC & Co. in 2009.

Before the ink was dry, I was contacted by a Mr. Hunter Barry. He told me his father, Craig, had been studying British firearms and British manufacturers for some years and had written various books on the subject. I contacted Craig, and we swapped stories and ideas. Before long we decided to team up and try to finish the job I had started with John Hopper.

Our initial efforts were overtaken by the story of little-known Irish manufacturer, Peter Tait of Limerick, and the late war uniforms his company provided to the Confederacy. In 2011, our short book on Tait was published: *Supplier to the Confederacy: Peter Tait & Co,*

Limerick. We then wrote our first major work, *Suppliers to the Confederacy: British Imported Arms & Accoutrements*.

Over the years Craig and I never stopped researching both SIC & Co. and Peter Tait & Co. and, with the success and critical acclaim of *"Suppliers to the Confederacy: British Imported Arms & Accoutrements,"* we thought it would be a good idea to combine both the SIC & Co. and Tait stories into one volume, rewrite most of it with new research and letters uncovered on Tait's involvement in the war effort and add new photos and appendices, which cover all the supplies of arms and other goods SIC & Co. had exported.

The result is *Suppliers to the Confederacy II: S. Isaac Campbell & Co, London/Peter Tait & Co, Limerick*.

David Burt, March 2014.

ACKNOWLEDGEMENTS

A special thank you to the staff of the South Carolina Confederate Relic Room & Museum in Columbia, South Carolina, especially to curator Krissy Dunn for allowing us access to the McRae Papers and the S. Isaac Campbell & Co. sub-series.

Special thanks also to Dr. Robert Jaffee for allowing us to use exclusive photographs of the Tait jacket of Private Hugh Lawson Duncan of the 39th Georgia Infantry.

Thanks to Bob McDonald for his help identifying the textiles used in the construction of the Duncan Tait jacket.

Thanks also go to, Chris Daly, Ron Field, The Museum of the Confederacy, The *Limerick Chronicle*, Limerick School of Art, Chris Becker (Maryland Historical Society), John E. Waite, Alan Thrower, Michael McComas, Larry Shields, Liam Dunne from the Limerick Studies, the Limerick City Council and Diane Cooke, among others in helping this book come to fruition.

PART ONE
S. ISAAC CAMPBELL & CO., LONDON

CHAPTER 1
SETTING THE STAGE

The Southern states seceded from the Union in early 1861 unprepared for war. They did not have the arms, accoutrements or uniforms necessary to field and equip large standing armies, and they needed all of it fast.

The antebellum strategy in most of the Southern states had been to secure whatever articles were necessary through importation rather than to attempt self-sufficiency by producing goods of their own manufacture. The Southern economy permitted this extravagance due to the continuing worldwide demand for Southern cotton and tobacco. When the Federal armories and arsenals in the Southern states fell into Confederate hands, precious few modern arms fell with them. While both sides were under-armed, the Federal armories in the South contained mostly obsolete flintlock and old style smoothbore muskets, and, at that, only about one third the total numbers on hand in the North.

Plans were underway by the Confederate Ordnance Department to begin the domestic manufacture of military equipment but attaining self-sufficiency was going to take time. Until then, most supplies would have to be imported.

England was an important source for military equipment. The British government, while officially neutral and politically opposed to slavery (having made the transition to an industrial economy), was sympathetic to the Southern Cause. Even so, the government established an official policy not to provide arms and equipment to the Confederacy as the French had done to the colonists during the Revolutionary War. However, the British government made no efforts to interfere with Confederate purchases from their own commercial military industrial complexes located around London and Birmingham.

On May 6, 1861, *The London Times Illustrated* ran a notice announcing the arrival of agents from both the North and South, which resulted in great excitement in the gun trade. The gunmakers, sensing their present advantage in the supply/demand equation,

Chapter One

were opportunistic about their pricing, and, as was noted at the time, "it does seem to me that we ought not to haggle too much...to save $10,000 might be to lose everything."[1]

Consider this example. After one Northern buyer made initial contact with John D. Goodman, who represented the Birmingham Small Arms Trade, he learned that 25,000 P-53 Enfields were immediately available at 60 shillings each (about £3 or $15.00).* However, before the deal could be concluded, another buyer identified as representing the Confederacy stepped in with an offer of 100 shillings each for the same lot of guns. Hence, the new price became 100 shillings ($25.00).

Business was conducted this way in the Gun Quarter, and the gunmakers willingly facilitated the fraternal slaughter on both sides without prejudice to politics. These were the halcyon days when Birmingham gunmakers lit cigars with £5 ($25.00) notes.

For their part, the American buyers were elbowing each other out of the way to procure as much military supplies as were currently available, as well as enter into new contracts for future deliveries. Complicating the tight supply, the British Volunteer movement was in full swing, and the local civilian units were purchasing their own uniforms and military arms. It was a very good time to be in the British gun trade.

★★★

Caleb Huse, Ordnance Purchasing Agent for the new Confederate government, arrived in London on May 10, 1861, just ahead of one group of Federal buyers, but ten days behind another group consisting of Francis Crowninshield and Thomas McFarland, who had been dispatched by Massachusetts. On board the same passenger ship, the *Persia*, was another agent representing New York. All three Northern agents were unaware of each other or their missions. Upon arrival, Crowninshield and McFarland both found a buyer already in London representing South Carolina.

Huse met with Elisha Fair of Alabama, former United States Ambassador to Belgium. Fair told Huse that all the large houses (factories) at Liege, Belgium, had more contract work than they could

---

*All conversion rates are British Sterling to Confederate Currency.

2

handle for several months, as he had recently inquired about obtaining arms for Alabama. In addition, the Liege gunmakers had the reputation of furnishing arms of diminished quality. Huse decided not to give any further attention to Belgium for the present.

Huse was, however, ready to do business on an enormous scale. He contracted with equipment brokers called commission houses that had connections in all corners of the worldwide trade. The London commission houses operated as agents for both the buyer and the seller, exacting a fee in the form of a percentage commission from each. One of the biggest commission houses in London was S. Isaac Campbell & Company (SIC & Co). They were anxious to do business with the Confederacy.

There was nothing that SIC & Co. could not provide, and Huse, lacking the time needed to scour the vast market himself, contracted with the commission house to buy massive quantities of small arms, accoutrements and ammunition desperately needed in the South.

Chapter 2
The Beginning

Samuel Isaac
(Courtesy of Bastien Gomperts)

"Early in 1861 Saul Isaac and his nephew Benjamin Hart, both of New York City, seeing the chance of financial gain, bought out the old and established military outfitting firm of S. Campbell & Co., 71 Jermyn Street, London. Thereafter, the firm operated as S. Isaac Campbell & Co."[1] Virtually, every word in this statement is incorrect, misleading or inaccurate.

The leading commission house in London was S. Isaac Campbell & Co. under managing directors, the brothers Samuel and Saul Isaac. Campbell appears to have been a silent or non-participating partner, who will be discussed in more detail later in the book.

Another long established myth that can be debunked is that the S in S. Isaac Campbell & Co. stood for Saul Isaac. The S actually belongs to Samuel, as he founded the company. Saul began working at the firm some years after SIC & Co. had been established.

4

Samuel Isaac was born in Chatham, Kent on November 22, 1812, the eldest of four children. Born to Jewish parents, Lewis Isaac (1788-1879), a furniture broker originally of Poole, Dorset, and Catherine Solomon (1789-1863), daughter of N. Solomon of Margate, Kent.

Isaac got his start in business as a perfumer, but by 1838, he was listed as an army and navy clothier. Before he was thirty, he owned two homes in succession in Chatham.

In 1836, Samuel's girlfriend, Isabella Simons, gave birth to their first child, Samuel Edward Henry Isaac – known as Henry. In 1841, the busy entrepreneur, still only twenty-five, finally married Isabella (twenty), and together, they had three more children, Lewis Henry (1841), Phoebe Grace (1842) and Michael Henry (1846).

Tragedy struck in 1842, when Lewis Henry died of water on the brain. In 1846, at the age of thirty-two, Isabella died of tuberculosis.

By 1845, Isaac was listed as a military tailor and outfitter with a business at 71 High Street, Chatham, Kent. He was also listed in the Post Office directory for Chatham as an army contractor, military tailor and outfitter, general East India passage agent, tobacconist, trunk manufacturer, china and glass dealer, general warehouseman and agent of the Church of England Insurance trust.

On June 21, 1848, Samuel married Emma Hart, the daughter of Stephen Hart. The union produced two children: Stephen Hart Isaac and Frances Isabel Isaac.

Isaac moved to London in 1848, where he opened his first office at 21 St. James Street, under the name of Isaac, Samuel Army Contractor. In 1849, the business changed its name to Isaac, Samuel, Army Contractor and Outfitter and by 1851, Samuel Isaac & Co., Army Contractors, Accoutrement Makers & General East India Merchants. In 1852, the name changed again. This time to the more recognized name of S. Isaac Campbell & Co., Army Contractors.

The company also kept its hand in the shoe-making business with a factory in Campbell Square, Northampton, in the heart of England. By 1856, the company was listed as providing boots and shoes to the British Army through the British Royal Military Depot or Weedon Barracks, which was just a short distance from Northampton.

★ ★ ★

On February 14, 1823, younger brother Saul was born in Chatham, Kent. He worked as a furniture manufacturer and cabinet-maker. By 1854, he had joined his elder brother at SIC & Co., where he was described in local directories as an army contractor. He was responsible for the financial affairs of the company.

On May 16, 1854, Saul married Emma Hart Isaac's younger sister, Miriam. They had two children: F.H. Isaac and Arthur Benjamin Isaac.

★ ★ ★

The traditional way for military outfitters to attract business was by having a prestigious London address. Historians have found it difficult in finding precisely when Samuel managed to find offices in London, but a document in the National Archives for Lincolnshire sheds light on when this happened:

> "...On May 28, 1860, a piece of land in Pall Mall Field or Saint James Field, and the dwelling house erected thereon, known as 71 Jermyn Street, Charles Worsley Anderson Pelham, Earl of Yarborough (sold) to Samuel Isaac, Saul Isaac and Charles Isaac (A cousin who left the firm shortly afterwards) of 71 Jermyn Street, Army Contractors."[2] In addition, on September 4, 1860, "alterations were made by Messrs Samuel Isaac, Saul Isaac and Charles Isaac to the premises at 71 Jermyn Street..."[3]

The 1861 census records the inhabitants of 71 Jermyn Street, London, as three clerks who worked at SIC & Co.: Henry Solomans, age fourteen; Henry Claridge, age eighteen and Lionel Hart, age eighteen.

★ ★ ★

The Isaac brothers were Jewish, so it is important to understand how they were perceived at this point in time in British history. The

Ashkenazi Jews, of which the Isaacs were descendants, were at best considered second class citizens with limited rights. In fact, only from mid-19th century onward were they even considered citizens at all, before then, they were tolerated inhabitants.[4] As Isaac D'Israeli noted in 1797, "This British land which, when the slave touches he becomes free, retains the child of Jacob in abject degradation. He cannot own the house in which he inhabits and is not able to elevate himself among his horde by professions which might ennoble his genius and dignify his people."[5]

Anti-Semitism prevailed throughout Europe at the time and could be summarized as follows: "The Jews are a very distant class of the inhabitants of London consisting of perhaps 20,000...though a few of them are respectable characters, the majority are notorious sharpers."[6]

British Jews were allowed to do business as brokers and many were successful in that line of work.

When the few rights contained in the Jew Bill (1753) were repealed, Sephardim Jews abandoned Judaism and either converted to the Church of England or practiced no religion at all. However, the Ashkenazim remained rabbinic, were confined to their insular communities and supported their synagogues in a more traditional way. Jewish emancipation in Britain dates from the passage of another bill in 1845 that permitted most rights of citizenship, except for election to Parliament.

CHAPTER 3
SHOE MAKING IN THE
MID-19TH CENTURY

As was so often the case during the Industrial Revolution, the centers of industry evolved around the sources of raw materials used in production. Birmingham was the center for gun making because of the proximity to iron ore in the Midlands. Likewise, shoemaking in Britain was somewhat centered around Northampton because of the proximity to the livestock necessary for shoe leather. Since the 1820s, Northampton had been supplying footwear throughout Great Britain, and although production was not yet mechanized and no factories existed, shoemaking was the major occupation.[1]

At the beginning of the 19th century, shoemaking was one of the largest single trades in England, right after carpentry. By 1861, there were 3,426 shoemakers in Northampton out of a population of 32,813, which equates to approximately 38% of the male population. Shoes from Northampton were exported as far as Australia and the United States.

Strong resistance to mechanization in Britain began in the early 19th century with the emergence of the Luddites. The Luddites perceived mechanization would deny the value of their craftsmanship and would eventually end their livelihood.

The Luddites took their name from Ned Ludd, who, upon the introduction of looms in the textile industry, supposedly broke two stocking frames with a hammer in a fit of rage. The Luddites wreaked havoc on the textile industry in 1811-12, even clashing with the British Army. This move sent a few Luddites to the gallows and a number of others to Van Diemen's Island.[2]

The Luddites were not the only ones who believed that mechanical devices separated the craftsman from the skills and eye and hand coordination essential to the craft.

Some of the finest footwear ever made was produced by skilled hands in the early to middle part of the 19th century, that is, before the wide spread use of machinery was introduced. Handmade boots and shoes were often more durable and more attractive than anything produced since, but not faster or less expensive.

Boot and shoemakers are now commonly called cobblers. However, the word cobbler was applied specifically to shoe repairmen. Those who actually hand-crafted footwear were known as cordwainers.[3] The term cordwainer is the Anglicization of the French word cordonnier, introduced into the English language after the Norman invasion of England in 1066. Cordwainer is derived from the city of Cordoba in the south of Spain, where the staple trades in the early Middle Ages were silver smithing and the production of cordovan leather, known as cordwain in England.

The London Trade Guild for this craft, which includes all fine leather makers (gloves, gilders, etc.), is the Worshipful Company of Cordwainers, twenty-seventh on the order of precedence of livery companies.[4] Livery companies are various historic trade associations almost all of which are known as the Worshipful Company of... with their relevant trade, craft or profession in the title. The medieval companies originally developed as guilds and were responsible for the regulation of their trades; controlling, for instance, wages and labor conditions.

The cordwainers were among the last trades to be affected by the Industrial Revolution, as sewing machines capable of penetrating thick leather were not patented until 1858[5], but the Northampton cordwainers did not wait until then to act.

In November 1857, the leading cordwainers held a meeting to discuss the introduction of machinery at the factory owned by Moses Phillip Manfield. During the meeting, Mr. Wilder, a shoemaker, identified that the purpose of the meeting was to "check the introduction of machinery, which was bound to bring ruin on them all."[6]

In April 1858, the Northampton Boot and Shoemakers Mutual Protection Society (shoemaker's union) was formed with the stated mission "to protect, raise and equalize wages."[7] A strike fund was set up and links coordinated with twenty principal workmen in neighboring Stafford. The line was drawn, and the issue contested in both Northampton and Stafford. Each side waited to see who would make the next move. It was not long in coming.[8]

In February 1859, twenty manufacturers of Northampton (and seventeen in Stafford) issued a statement confirming the cordwainers worst fears: Sewing machines were coming to both cities, and it was matter-of-factly announced in the following statement:

> *"That in consequence of sewing machines being extensively used in the cities and principal towns in the United Kingdom, so as seriously to affect the demand upon the wholesale houses, any further delay in the introduction of them, by the manufacturers of Northampton, would be permanently injurious to the interest of the trade generally. And in accordance with this conviction, it was decided to introduce the machine sewn tops simultaneously into their respective trades."*[9]

The reaction of the Mutual Protection Societies was to call a strike, urging as many shoemakers as possible to leave Northampton and seek work elsewhere, as long it was not in Stafford. The strike failed to rouse the support of many workers because Northampton shoemakers did not have any objection to the introduction of sewing machines, so long as the machines did not threaten any jobs. Those who did strike went to Leicester and took jobs on sewing machines in the factories there.

After the strike ended in May 1869 and just as business returned to normal, the construction of SIC & Co.'s new state-of-the-art factory on the corner of Campbell Square and Victoria Street, Northampton, was completed. It sat next door to the second oldest shoe factory in England.

Samuel had been quick to grasp that the principles of mass production were the only way forward. There was no need for a lengthy apprenticeship. Factory workers could be hired beginning at the age of nine years. The sewing machines required a level of dexterity which could be learned in a few months; hence the skills of the old trade were not necessary or even desirable. The plan was, of course, contested by the Northamptonshire Mutual Protection Society. On March 12, 1859, the *Northampton Mercury* asked (and answered) the following question:

"Is it not the factory system which is contemplated by employers more so than the machines? Yes shop-makers, it is the infernal factory system they want to introduce."[10]

SIC & Co. placed an appeal for workers in the *Northampton Mercury*, which read in part:

"To the boot and shoemakers of Northampton.

"You live by work. We want work done on fair terms and for fair wages. That being so, our object is to establish those proper and just relations, which should exist between employers and the employed. We have built, at great cost, extensive premises in which to carry on the manufacture of boots and shoes.

"They are arranged upon the best plan. The rooms are large, lofty and well ventilated, and kept warmed at a uniform, moderate and healthy heat by nearly two miles of hot water piping. The engagements will be permanent for all those who are willing to do so, each day, a good day's work under the superintendence of our foreman.

"The work will all be piece work. The attendance must, for your sakes as well as ours, be regular. The hours fixed are – in summer from 6 to 8, from 8 1/2 to 12, and from 1 to 6 o'clock; and in winter from 8 to 12, and from 1 to 4, and from 4 1/2 to 8. (Both shifts were 10-1/2 hours long.)

"We intend to employ machinery. We state that plainly, because we know that many of you have striven against the introduction of machinery, but we submit to you, we are glad to know that many of you are aware of the fact, that machinery must be employed and to struggle against it is to fight with science, and an attempt to put a stop to the progress of the human mind.

"We intend to employ women and children on the premises. Some of you have objected to that being done; but it is obvious that those women who work at

11

machinery must be employed on the premises. For them, separate work rooms, entrances, stair cases and personal accommodations have been provided; and they will be superintended entirely by females. Four men will work at each table. The men at each three of the tables may elect from among themselves an overseer, who will see that the work is properly done.

"We have heard your objections to what is called 'the factory system.' We submit to you that the system that we propose is not the 'factory system.' It is a carefully considered system of constant, orderly regulated work, without any of the bad features, which have made the factory system distasteful to you; for example –

"Married women may have work at home;

"Parents may bring their children as apprentices;

"Men and women will be kept separate;

"Workmen will be allowed to choose their own overseers;

"Subdivision of labor will not be attempted.

"Instead of being obliged to work in the close, confined rooms of your cottages, you will labour in healthy, commodious and well ventilated apartments. In regular hours of orderly labour, free from domestic hindrances, you will be able to do more work and earn more money in less time than you can now.

"Your children employed under a well-regulated system will require habits of industry and order and become more valuable to you. Regular half-holidays will afford you opportunities for amusement and recreation.

"We submit these important changes to you in all frankness, and in the hope and belief that you will see their reasonableness and advantage."[11]

However, the thought of fixed hours represented an abrupt end to the traditional autonomy enjoyed by craftsmen, and it was precisely this attempt to introduce both machinery and the factory system which provoked the earlier strike. In addition, the Isaacs only intended to employ non-society workers and children, meaning those who did not support the strike of 1858-59.

A cordwainer and his family work from home.
(Courtesy of The Honourable Cordwainers Company)

Not wishing to provoke any confrontation with the Protection Society or the local shoe makers, the Isaacs decided to opt out of manufacturing entirely and lease the factory to Turner Bros, Hyde & Co. in 1861. The Isaacs eventually sold the factory outright to the Turners some years later. But it is strongly believed that the Isaacs still held an interest in the factory, although they hid their ownership. It is also thought that Hyde may have been Samuel Isaac, although this fact was never conclusively proved.

Richard Turner, whose shoemaking firm was founded in the 1840s, recognized the economic advantages of the factory system but decided not to adopt it for fear of antagonizing the labor force.[12] Charles Parker, son of a bootmaker, who ran the factory when the Isaacs owned it, stayed on as general manager after Turner Bros took over.

By 1862, the operation employed 300 people, and it was soon mass producing 100,000 pairs of shoes and boots per year.[13] By 1865, the factory had huge markets in Australia and New Zealand.[14]

SIC & Co. contracted with Turner Bros for military footwear. Hence, a portion of the army shoes (and boots) from the same factory that SIC & Co. built was later brokered by them for shipment overseas on blockade runners and worn by Confederates on the battlefield.

Chapter Three

The Isaacs' decision to sell the shoe factory served to demonstrate the process by which the American plan of mass production, or factory system, was grudgingly adopted by British workers. This was apparently the case in London as well as Northampton and Stafford, as only one new firm there was recorded as utilizing sewing machines in the shoe industry before 1859. The shoemakers' strike of 1858-59 stands as the last labor strike to occur as an outright attempt to prevent mechanization.

CHAPTER 4
WHO WAS CAMPBELL?

It has been a subject of mystery for some time as to the identity of the Campbell in SIC & Co. Evidence suggests that Dugald Forbes Campbell was the Campbell in question.

Campbell was born on October 19, 1814, in Glasgow, Scotland. By 1847, he was assistant manager of the Colonial Bank in London. He also translated French language books into English. Two of the works he translated were *History of the Consulate and the Empire of France under Napoleon* by M.A Thiers in 1851 and *Remarks on the Production of Precious Metals and the Depreciation of Gold* by Michel Chevalier in 1853. He was on best terms with Prince Louis Bonaparte, III of France.

On January 16, 1862, Campbell wrote to the Paris correspondent of the *London Morning Post* about the Charleston harbor grievance.

> "My Dear Brown:
>
> "How comes it that you have never alluded in your correspondence to the Yankee doings in Charleston harbour and the indignation threat roused in France? Upon inquiry you will find that fully three weeks ago, France and England, separately, addressed the strongest possible remonstrances to the government at Washington against the vandal like act, then in contemplation. It has been consummated in spite of our remonstrances.
>
> "The foregoing I give you for a fact. I learn further from an excellent quarter that instructions have gone to M. Mercier to notify the Washington government that France can no longer recognize the blockade of the Southern port, that the blocking up of the harbor of Charleston was uncalled for had the blockade been 'effective.' England approves of this and will back up France. The lead however will, on the present occasion, be taken by the Emperor.

"It is said too that H.M., will in his speech on the 27th instant, denounce the barbarous mode of warfare adopted by the North and proclaim the blockade no longer binding on France. What joy such an announcement will occasion in Manchester and other places now sorely tried by the cotton famine?

"The enclosed from the Herald of 6th inst. is the programme of the Conservative party on the American question. The party can marshal 314 men, at a division, and as 127 liberals and radicals (some of them good speakers and men of weight) are pledged to support a motion for the immediate recognition of the Confederate States and the raising of the paper blockade, the Ministry will be beaten if they do not make a virtue of a necessity and anticipate the action of Parliament. The motion in question will be made and seconded by advanced liberals and supported by the conservatives 'en masse.' Make what use you like of the preceding.

"Do you know whether M. Fould has determined to raise a loan? If you do and can give me the figure and times privately by Monday morning's post, the information might put something into both our pockets. Of course you have seen Sir Robert M.P. and heard his 'veni, vidi, vici.' Was it he who pitched into Lord Cowley so hard, the other morning in the Times?

"Yours Sincerely,

"D. Forbes Campbell"[1]

In another correspondence, Campbell denounced the blocking of the entrance of Charleston Harbor with sunken ships as barbarous, said the war was waged by the Northern states for political and territorial dominion and that the extinction or limitation of slavery with them was of an altogether secondary consideration. The U.S. Consul in Paris, John Bigelow, noted in his dispatch, "(that) the reader may expect to hear again of this Mr. Campbell."[2]

Other events linking Campbell to SIC & Co. came in another letter written by Bigelow. He noted in autumn 1865:

> *"I was the guest at a costume ball in Paris when I was presented to an English gentleman. The gentleman's name was Dugald Forbes Campbell. It appeared during the course of our interview that he was acting as an attorney for S. Isaac Campbell & Co. of London, the owners of the barque called the Springbok, which had been overtaken on our coast and condemned as a blockade runner."[3]*

Another supporting piece of evidence that makes it clear that Campbell was not a participating member of the firm comes from post-bellum legal documents filed on behalf of SIC & Co. to reclaim the same illegally seized *Springbok*. It reads:

> *"The claimants in the prize court of the cargo of the Springbok viz the firm of S. Isaac Campbell and Co. of London and Thomas Stirling Begbie also of London are the memorialists here. The firm of S. Isaac Campbell & Co. was at the time of these transactions composed of Samuel Isaac and Saul Isaac and had no other partner.*
> *"The duly accredited attorney in fact of the memorialists before this commission Dugald Forbes Campbell, Esq. of London, whose powers duly verified are filed with the Commission, is not to be taken from the name of Campbell appearing in the firm of S. Isaac Campbell & Co. to have had any connection with the transactions of the voyage of the Springbok. That firm had no partner of the name of Campbell as is shown in the prize causes and in the present memorial. Mr. D. Forbes Campbell represents the existing interests in the claim, which, as is stated in the memorial, are largely those of creditors of the original parties."[4]*

It is probable that Campbell received a silent partnership either via outright purchase of a partner's interest, or from the loan of a principal sum to Samuel Isaac to expand the firm in 1852 because that is the year the firm changed its name from Samuel Isaac, Army Contractors, Accoutrement Makers & General East India Merchants, to S. Isaac Campbell & Co. Army Contractors.

During the same period, Campbell was involved with literary translation and legal work. However, Campbell lists himself as agent for SIC & Co. when he bought £30,000 ($150,000) worth of Erlanger Bonds in 1863.

Campbell seems to be in the picture, but always at arm's length. Yet, who better to have as an agent or silent partner in SIC & Co. than a man of high status, influence and wealth, especially when you are seeking to secure military supply contracts with the British military?

Campbell died in 1886, the same year Samuel Isaac passed away.

Weedon Bec Royal Ordnance Depot, Northamptonshire
(Courtesy of Subterannea Britannica)

Chapter 5
The Weedon Bec Scandal

O n February 1, 1856, the *Jewish Chronicle* (London) reported the following:

> "The contract for the supply of clothing to the whole of the British Army in the East has been taken by Messrs Isaacs, the Army contractor of Chatham and St. James St., London. It is anticipated that this firm will also have the contract for providing clothing and regimental necessaries to the whole of the Army, the government having decided on placing a contract in the hands of one contractor only."[1]

★ ★ ★

The Royal Ordnance Depot, as it became known, was first conceived in 1803. An act of Parliament allowed for the purchase of 150 acres of land in Weedon Bec, a village in Northamptonshire situated right in the heart of England. The site was chosen because of its proximity to the canal network and the realization that storage of military supplies near the coast was no longer prudent because of the threat of invasion by France. In fact, Northamptonshire and the depot were thought to be so safe that plans called for the Royal family to be sent there by canal from London if Napoleon were to invade the country.

The depot stretched out along the Nene valley above the village. A barrack built to hold 500 men overlooked the depot to the north, close to the Coventry Road. The depot became the main clothing and general stores for the British Army. Initially, the depot had eight storehouses and four magazines.

The storehouses were of brick construction and faced with stone, each two storey high. They were 160'-0" long and 35'-0" wide and divided into four rooms. One of the buildings was converted into a military prison which contained 121 cells. The adjoining buildings were used as a hospital and chapel. The eight buildings covered approximately a quarter of a mile. The magazine buildings were 300 yards to the west in a separate walled enclosure. As early as 1809, the depot was used for storage and to issue small arms and ordnance.

In order to move goods quickly into the depot, a tributary was constructed from the nearby Grand Junction Canal, which ran between the two rows of storehouses. At each end of the main enclosure, two lodges were built over the canal, each equipped with a movable portcullis. Cupolas surmounted each of the lodges. On the east lodge, the clock still chimes today. The canal cut continued into the magazine, passing through a smaller building and portcullis. At the western end, a fourth portcullis led to a barge turning area outside the perimeter wall. Barges were also able to turn in a canal basin within the magazine enclosure.

Gunpowder was delivered to the depot by barge, where it was packed into barrels and boxes and re-issued. The canals remained in operation until the standardization of the railroad gauges. Then supplies were shipped to Weedon Bec via the railways.

The Main Canal
(Courtesy of Subterranea Britannica)

James Elliott was the principal storekeeper at the Weedon Bec Depot, having previously served in the Ordnance Department in London. He was appointed to a commission to inquire into the military expenditure in Canada from 1837-39. In 1845, he was promoted to chief clerk at headquarters in Canada. In 1851, he was appointed a member of the commission to inquire into naval and military establishments abroad. In 1855, he received the appointment of chief superintendent of military stores at the Weedon Bec Depot. He was regarded as a shrewd, intelligent and upright public servant.

Things change. By December 1857, Elliott was suspected of irregularities in his depot's accounts and neglecting his duties by spending more and more time in London.

The government determined to remove him to Ireland. Elliott was discharged from the service, but was retained at the depot to balance his cash and stock books. He remained at Weedon Bec until May 1858, when he decamped to America to avoid prosecution.

Elliott's difficulties arose when he received monies from the government to pay accounts. He embezzled the funds and was unable to meet additional accounts, including wages due to be paid at the depot. He applied to the War Office for money to cover the wages for the month of May but was refused. By May 14, he was unable to pay anyone.

Samuel Isaac was at Weedon Bec a few days later, and Elliott called on him. Elliott asked Isaac for a loan of £500 for three days. Isaac agreed without asking Elliott why he needed the money. The money was paid into the Northampton Bank on May 16, and the following Monday, Elliott gave a check to one of the clerks at the depot to cover wages.

The money given by Isaac was still not sufficient to cover the outstanding debts, so Elliott borrowed an additional £350 from Cox & Co., a firm commonly called the Bankers of the British Army.[2] Elliott, with the funds from Cox & Co. in his pocket, embarked for America. During the War Office's investigation into Elliott's misconduct, the committee members discovered Isaac's check in the account and found that a balance of the sum remained.

On November 19, 1858, Isaac was summoned to appear before a Royal Commission. He was questioned about the loan to Elliott. At the same time, Isaac was accused of re-selling, at higher prices, inferior and rejected kits to the depot at Chatham.

At the time of the inquiry, SIC & Co. had two contracts with the War Office. The first contract was for clothing and regimental necessaries, which included army boots, while the second contract was for regulation boots. The commission alleged that Isaac transferred inferior boots from one contract to the other to save himself from being fined for late deliveries and poor workmanship.

The *London Examiner* reported:

> "The difference between Weedon and Chatham – it appears that one of the largest army contractors contracted to supply soldiers' kits at £2,11s.3d. each, and, his contract having been accepted, several thousand kits were sent into the clothing stores at Weedon, when it was discovered that the articles were of an inferior description and not worth the sum paid for them by the government. On this discovery being made, the kits were returned to the contractor, who has since supplied the same kits to the troops at Chatham for £3,8s.9d. each. It thus appears that articles rejected by the government as not being worth £2,11s.3d. at Weedon, are considered to be worth £3,8s.9d at Chatham and are purchased by the government for the troops at that price.
>
> "With regard to knapsacks, some supplied by Isaac were of inferior quality, the canvas being rough, and the corners not being waterproof.
>
> "There was also confusion over the name Messrs Isaac & Campbell being marked on them; there had been confusion over whether the knapsacks which had failed a test in which samples were compared against a sealed pattern were genuinely from the same supplier as that of the sealed sample."[3]

When asked why he had loaned Elliott £500, Isaac said that he made the loan to a friend, after having been told that there was not enough money to pay wages due. It was a loan to be repaid in three days.[4]

Chapter Five

Elliott wrote Isaac from his home in New York offering to back up Isaac's claim that the money was a loan. This letter was read to the commission:

"*New York, August 30, 1858,*

"*Dear Sir – Having read in the papers a statement referring to the loan of £500 made by you to me, and to various public dealings with you as a contractor, which statement, if explained, might give to these transactions character not in accordance with the facts, I think it right, both in justice to myself and to you, to say that, as to the £500 having been given as a bribe, or for any favors shown to you in the discharge of my public duties, the assertion is a cowardly and calamitous falsehood, which no man, however high his position, would dare to advance were I in England. The certificates for stores not delivered must, to any man conversant with the practice of the service, be an obvious impossibility. The certificate, previous to receiving my signature, must of necessity have received those of the inspectors who examined the stores both as to quantity and quality, and upon the faith of those signatures would have been attached, and without those, notwithstanding my signature, the Accountant-General would not allow the claim for payment.*

"*In no way and on no occasion have I favoured or offered facilities to your firm which were not extended equally to all other contractors, among these are many men of high honour and character; to them I may confidently appeal as to whether I have not on all occasions endeavoured to the utmost of the meager assistance sparingly afforded to me, to ensure them promptitude of payment and a fair inspection of their supplies.*

"*That I ever allowed you or any other contractor to suffer by my neglect of duty or absence from Weedon, I utterly and emphatically deny. No man ever made himself more a slave to the public service than I did. Had*

I been less zealous, I might have been less maligned by those who deem it honourable to cover their own defects at the expense of an absent and oppressed man who, had descended to what they would attribute to him during the past twenty years of his serfdom, possessed ample opportunity for becoming as wealthy and adulated as to be now poor and unfriended.

"I regret my inability at present to repay you the £500 but shall esteem it and first lieu upon whatever balance is due to me from the government arising out of a claim which I am now prosecuting exceeding the amount against me by the War Office, and which however at present opposed, I believe to be irresistible both in law and equity.

"I may add that I am ready at any moment to declare, on oath, that the loan of £500 was the first and only private monetary transaction of any description between us, which, at the time of borrowing, I was fully persuaded it would have been in my power to replace out of money that I then had a prospect of raising.

"I remain, dear Sir, yours truly,
"J. Elliott"[5]

Isaac was cross examined by the commission on November 22, 1858. He was asked about the circumstances of the loan. He testified that he did not know what exactly Elliott wanted the money for, and he did not enquire as he found Elliott to be a man of high character.

Not satisfied with Isaac's answer, the commission pressed him further. Isaac responded:

"Mr. Elliott was a man I had seen every week for three or four years, never supposing him to be in difficulties. I assure you that I had the very highest opinion of Mr. Elliott. I confess it was a very foolish act."[6]

Elliott failed to supervise the clerks, which led to the clerks not properly recording which deliveries belonged to which contracts. As a result of the poor record keeping and the failure of key witnesses to provide evidence against SIC & Co. or Samuel Isaac, the investigating committee could not prove that either SIC & Co. or Isaac were guilty of malfeasance. The official Parliamentary report concluded:

> *"Part of the problem seemed to be that the kits, supplied by Isaacs contained boots, and the boots themselves were contracted items in their own right; if boots that were part of the kits failed an inspection, boots from the other contract could, for obvious reasons be substituted. The complexity of recording these transactions were not carefully recorded, all concerned seemed to get into a mess when asked to account for them."[7]*

However, the commission decided that the loan was seen as an improper gratuity, which was technically a breach of contract. Consequently, the SIC & Co. contracts were cancelled immediately, and the firm was banned from any future contracts with the War Office.

A letter from the War Office to SIC & Co. read:

> *"Sirs, I am directed by Major Gen Peel to transmit a notice to you terminating your contract of 1st May 1858, for the supply of army regulation boots. General Peel desires me to observe to you one of the terms of the contract is that 'it be declared void should the contractor pay any gratuity or reward to any person in the employment of the War Office,' and that in reference thereto, his attention has been called to the fact that on the 15th day of May last, your Samuel Isaac paid to Mr. J.S. Elliott the sum of five hundred pounds. General Peel is aware that the transaction is stated to have been a loan, but if so, it was (as it appears from your Mr. Isaac's statement) made under such easy circumstances to Mr. Elliott, that the Secretary of War can only consider it in the nature of a gift or gratuity, and which, as you are aware, every officer in the establishment is precluded from accepting. General Peel, therefore*

cannot consent to your continuing any longer one of Her Majesty's contractors for this department."[8]

It seems the War Office was considerably less incensed with Cox & Co. than with SIC & Co. over their respective loan(s) to Elliott. The real issue was not the loan to Elliott, but rather the discovery of the Isaacs' transferring boots from one contract to another, which, although not proven, was still believed to be the truth.

The loss of the War Office contracts came as a shock to Isaac, who, over the following months, sent several letters to have SIC & Co.'s contracts reinstated.

⋆ ⋆ ⋆

On January 15, 1858, Felicie Orsini attempted to assassinate Napoleon III, in what became known as the Orsini Affair. Orsini had traveled to England and arranged for a fellow conspirator to obtain the bombs from a firm in Birmingham. Tensions rose between Britain and France. With one third of its army already involved in foreign wars, Britain's military defenses had been stretched thin. The government called for 160,000 civilians to form a new volunteer force to defend the homeland. These new units would be responsible for outfitting themselves.

Being former suppliers to the British military, SIC & Co. still had sub-contractors in place, so the company was perfectly placed to provide supplies for this new market. Purchases from the volunteers kept the company in business while Isaac sought reinstatement with the British Military.

CHAPTER 6
SIC & CO. BOUNCES BACK:
SUPPLYING THE VOLUNTEERS

The original British volunteer units were formed during the Napoleonic-era with the stated purpose of defending the homeland from France, which had forces massing on the coast near Boulogne. These volunteers were strictly temporary, raised in a hurry to meet that specific threat and were never part of the regular British Army. For example, a volunteer's idea of military duty would not include fighting overseas but defending the homeland in case of invasion. Over 460,000 volunteers enlisted from 1794 to 1814. After the British victory at Trafalgar, the volunteers were not needed and eventually dissipated.

However, on the heels of the Crimean War and the Sepoy Rebellion in India a few years later, the British Army was spread thin across the Empire. Less than a third of all effectives remained in England. When Great Britain was implicated in the Orsini Affair, England was rife with rumors of a French invasion.

In what became known as The Panic of 1859, the British government authorized the formation of volunteer and artillery corps. Many communities had rifle clubs, and these members were encouraged to enlist under the following regulations:

- ★ The corps was only to be formed on the recommendation of the county's lord-lieutenant.
- ★ Officers were to hold their commissions from the lord-lieutenant.
- ★ Members of the corps were to swear an oath of allegiance before a justice of the peace, deputy lieutenant or commissioned officer of the corps.
- ★ The force was liable to be called out "in case of actual invasion, or of appearance of an enemy in force on the coast, or in case of rebellion arising in either of these emergencies."[1]

* While under arms, volunteers were subject to military law and were entitled to be billeted and receive regular army pay.
* Members were not permitted to quit the force during actual military service, and, at other times, had to give fourteen days notice before being permitted to leave the corps.
* Members were to be returned as effective if they had attended eight days of drill and exercise in four months or twenty-four days within a year.
* The members of the corps were to provide their own arms and equipment and were to defray all costs except when assembled for actual service.
* Volunteers were also permitted to choose the design of their uniforms, subject to the lord-lieutenant's approval.
* Although volunteers were to pay for their own firearms, the guns were to be provided under the superintendence of the War Office so as to ensure uniformity of gauge.
* The number of officers and private men in each county and corps was to be settled by the War Office, based on the lord-lieutenant's recommendation.

Originally, corps were to consist of approximately 100 men of all ranks under the command of a captain, with some localities having subdivisions of thirty men under a lieutenant. The purpose of the rifle corps was to harass the invading enemy's flanks, while the artillery corps was to man coastal guns and forts.

According to regulations, 25% of the men were supplied arms by the government for instructional purposes. In case the units were called out for actual service in the field, the whole force would be armed by the government. Some volunteer units did eventually see action in the *fin de siècle*[2] Boer War but were using much more modern weapons by that time.

By January 1860, there were 6,716 men in the volunteer artillery brigades and 67,078 men in the volunteer rifle units.[3]

British regulars looked down upon the volunteers, at best, as parade fodder and, at worst, as worthless. They were lampooned in cartoons depicting a festively plump man in a gaudy uniform under the title *John Bull Guards His Pudding*. The volunteers were also chastised in song, and the following ditty comes from the British magazine *Punch*:

"Some prate of patriotism and some of cheap defense,
"But to the high official mind that's all absurd pretence,
"For all the joys of snubbing, there's none to it so dear,
"As to snub, snub, snub, snub the British Volunteer..."[4]

★ ★ ★

SIC & Co. applied to the War Office to supply sealed patterns for uniforms and equipment for this rapidly growing new force. An 1860 memo gave a list of the items the commission house would supply.

★ *"Tunic complete, pair of trousers, forage cap, set of patent accoutrements, waist and shoulder belt, pouch, ball bag, cap pocket and bayonet frog complete £2,15s.9d.*
★ *"Overcoat is from 30 shillings.*
★ *"These are the wholesale prices on which Messrs Isaac & Campbell & Co. of St. James Street, the army contractors who supplied the sealed patterns to the War Office, are willing to contract for the equipment of the Corps."*[5]

Upon being awarded a contract for uniforms and accoutrements, both Isaacs became involved in the volunteer brigades. On March 3, 1860, Samuel Isaac created his own unit and was commissioned a captain commandant of the 5th Northamptonshire 1-A Battalion Rifle Volunteers. On July 24, 1868, Saul Isaac was elected captain of the 46th Westminster Rifles.[6]

The men in the Northampton Volunteers were mostly workers from the Isaacs' former shoe factory. The officer corps came from men of the most prominent families in Northamptonshire. The volunteer companies drilled several times per week.

An article in the *Jewish Chronicle* told the story:

"Messrs Isaac Campbell & Co.'s Rifle Brigade.

"From the 'Northampton Herald,' we learn that a rifle brigade has been formed in Northampton, chiefly consisting of persons in employ of the above firm. Their number is about eighty; and their commanding officer is

Captain Isaac. Having been sworn in at the Guildhall, the mayor addressed the volunteers at some length on the nature of the new duties which they had now undertaken. Capt Isaac replied in an appropriate speech, in which he impressively pointed out to the men how they should make themselves useful to their country and their native town, their families and themselves. The brigade then marched to a hotel where a festive evening was spent, various patriotic toasts being proposed. We must add that the uniform, approved by the Lord Lieutenant, is supplied free of cost to the volunteers of the firm."[7]

From 1859 on, SIC & Co. would become the sole provider of uniforms and equipment to eighteen of the newly created volunteer rifle corps regiments. The volunteer regulations for accoutrements were listed as:

"Accoutrements, to be provided at the expense of the members will consist of waist-belt of black or brown leather, sliding frog for the bayonet, ball-bag containing cap-pocket and twenty round pouch."[8]

The accoutrements supplied to the Confederacy by SIC & Co. and which have turned up over the years, started out as volunteer corps equipment. One item, supplied in large quantities, was the ball bag. It differed from the regular army bag because it had a cap pocket on the inside. A belt was shipped to the Confederacy in the 1854 army pattern with a plethora of belt buckle designs. The belt came in brown or black leather.

Accoutrements were often not of the best quality. Items were constructed from cheap leather and poor craftsmanship. The items did not have to pass inspection as the British Army patterns had to. Neither did the items have to stand up to the rigors in the field.

☆ ☆ ☆

In the spring of 1861, General P.G.T. Beauregard, commanding the Confederate batteries in Charleston, opened fire on Federal forces

Chapter Six

in Fort Sumter. For SIC & Co., struggling to survive, this war across the ocean would bring much needed business.

CHAPTER 7
COLONEL JOSIAH GORGAS, CSA

Josiah Gorgas
(Courtesy of the Library of Congress)

The man chosen by the Confederacy to lead the new Ordnance Department was a Northerner. Josiah Gorgas was born in Running Pumps, Pennsylvania, on July 1, 1818, one of ten children. His father, Joseph, was, at various times, employed as a farmer, clockmaker, mechanic and innkeeper. Because the family was so large, Joseph and his wife struggled to provide for its needs. Hence, Josiah did not have an opportunity to obtain a formal education.

One distinguishing physical feature staring back from surviving photographs of Gorgas is the slightly crooked nose, which he earned when he tripped and fell at the age of three.

Gorgas left home at seventeen to live with an older, married sister in Lyons, New York. He found employment in a newspaper printer's office and came to the attention of Graham Chapin, the company's lawyer and the district's congressman. Chapin opened his law office to Gorgas, who took advantage of the invitation to study the law.

Chapin obtained an appointment for Gorgas to the United States Military Academy. Gorgas graduated in 1841, sixth in a class of fifty-two. He was assigned to the Ordnance Department and served under General Winfield Scott during the Mexican War.

After the war, Gorgas was assigned to a post in Mt. Vernon, Alabama. He met and married Amelia Gayle, the daughter of former Governor of Alabama, John Gayle. The couple had six children.

Gorgas and his family were living in Philadelphia when the first Southern states seceded in the winter of 1860–61. In February, he was offered a commission in the Confederate Army, but declined the offer. When he learned that he would be transferred from the Frankford Arsenal to foundry duty under Benjamin Huger, he resigned from the United States Army effective April 3, 1861.

General Beauregard recognized Gorgas' organizational talent and recommended him to head the Ordnance Department. Jefferson Davis accepted Beauregard's recommendation and issued Special Orders Number 17, Confederate Adjutant General's Office, officially assigning Gorgas as Chief of the Bureau of Ordnance, effective April 8, 1861.

When Gorgas arrived in Richmond, none of the supply bureaus had anything to issue to the armies now being assembled. In addition, there were limited funds available for purchasing imported military wares. Until Gorgas could build the necessary infrastructure to produce supplies, he had to organize a provisional operation to meet the immediate needs of the Confederacy.

One of his primary tasks was to select a purchasing agent to travel to Europe and obtain arms and munitions. He selected Caleb Huse.

On April 15, 1861, Adjutant and Inspector General Samuel Cooper issued orders to Huse, which read in part:

> "You are hereby directed to proceed to Europe, without unnecessary delay, as the agent of this government, for the purchase of ordnance, arms, equipments and military stores for its use. Detailed instructions as to the nature and extent of those purchases and as to their shipment, with a view to

speedy and safe transit, will be given to you by the chief of the Bureau of Ordnance."[1]

Several months later, Huse's orders were changed to include the purchase of small arms of any description. On August 2, 1861, Secretary of War L.P. Walker summed up the situation in a letter to Huse:

> *"This Department wishes you to consider your original instructions as no longer binding in their strict sense, but empowers you to construe them liberally in the line of objects to be obtained. To meet the large forces our enemy is endeavouring to hurl against us, we must have additional arms before supplies can be obtained from our factories, just now going into operation. Operate with a free hand to meet our pressing needs and ship safely and consider your credit extended to the full of this demand."*[2]

At the behest of Gorgas, the War Department set up a priority system which was to be adhered to at all times. Small arms and powder were placed at the top of the list, and Huse was instructed to obtain these items to the neglect of all other stores. To pay for these purchases, the Treasury Department made payments in cotton. Even so, Gorgas had to fight for every penny he sent to Huse, whose shopping list grew ever longer.

CHAPTER 8
CAPTAIN CALEB HUSE, CSA

Caleb Huse
(Courtesy of The Supplies for the Confederate Army)

Caleb Huse was born in Newburyport, Massachusetts in 1831. He enrolled in the United States Military Academy at the age of sixteen and graduated in 1851, seventh in his class. After graduation, he remained at West Point and taught chemistry, mineralogy and geology.

In September 1860, he took a leave of absence to accept the position of Superintendent, Professor of Chemistry and Commandant of Cadets at the University of Alabama. His appointment caused some consternation for many of the students, who did not appreciate his disciplinary techniques (for which he had been hired) or his Northern roots.

In April, Huse was offered a commission in the Confederate Army. He accepted the offer and backdated his letter of resignation from the United States Army effective February 25, 1861. Huse was excoriated in the Northern press as a traitor who had "...abandoned his state, country, principles and friends to engage in the business of furnishing supplies to the rebels..."[1]

Huse's Northern roots aroused suspicion in the Confederate government. Jefferson Davis ordered Major Edward C. Anderson, a planter from Savannah, to travel to London to examine Huse's conduct and to replace him if he was found to be disloyal.

On May 18, 1861, Walker wrote to Anderson:

> "You are hereby authorized, should circumstances in your opinion demand it, to supersede Capt Caleb Huse, who was sent to Europe as an agent of this Department to purchase ordnance, arms and munitions of war and to take possession of any assets or credits placed to his account as such agent."[2]

Anderson arrived in England in June. He found Huse to be completely competent and loyal to the Confederate cause. On July 17, 1861, he reported to Richmond:

> "My first duty on my arrival in England was to comply with the instructions which I had received in Montgomery and to scrutinize very closely the operations & sentiments of Capt Huse. To this end I conferred very fully with Mr. Prioleau (Charles Kuhn Prioleau), our financial agent, & more particularly with Captain Bulloch (James Dunwoody Bulloch), whose closer intimacy with H. would enable him to afford me correct information. From both, I received the most satisfactory assurances of the fidelity & loyalty of Mr. H."[3]

Anderson, satisfied with the Huse's performance, turned his attention to purchasing supplies.

Chapter 9
Anderson, Huse and SIC & Co.
Begin to Supply the Confederate
States

What was a commission house and how did it work? It worked like this: A wishes to import goods from B, but because A is not a recognized government, A cannot establish credit. The commission house steps in as an intermediary, finds and contracts for the goods from multiple parties and often buys in bulk directly from the manufacturers. The commission house arranges credit or financing for A. For providing these services, the house charges a fee of 2% to 2-1/2%.

During the war years, London was ideally suited for the commission houses. The support services and expertise in international trade, foreign exchange brokerage, financing and communications were all available. London provided the 19th century equivalent of international one-stop shopping.[1] Wars large and small were erupting in Brazil, New Zealand, Italy and Mexico. Many foreign governments brought British military supplies during this time.

★★★

Fraser, Trenholm & Co. was founded in Charleston, South Carolina as a trading and finance firm and importer/exporter for plantation owners and businesses. By 1861, the firm had acquired two subsidiaries: Trenholm Brothers in New York and Fraser, Trenholm & Co. in Liverpool.

The secession of Southern states opened a door for the company to do greater business. The New York office was closed, and two new offices were opened in Nassau and Bermuda. The company's Liverpool office soon became the connection for the Confederacy's naval and financial institutions in Europe.

At the beginning of the war, the Confederate government deposited $500,000 with Fraser, Trenholm & Co. in Charleston and provided letters of credit to purchasing agents to purchase and ship necessary supplies. Upon his arrival in England, Huse went to the Liverpool office. He wrote of the initial meeting.

"I arrived in Liverpool on the 10th of May, and, at once, put myself in communication with the house of Fraser, Trenholm & Co., on whom I had letters of credit. I found these gentlemen, and especially Mr. Prioleau, a member of the firm, ready to do everything in their power to assist me in carrying out successfully the object of my mission. On presenting my letters, it appeared that I had actually but £10,000 with which to purchase arms, etc. The letter from the Secretary of the Treasury to Messrs Fraser, Trenholm & Co., informing them of my drafts on the C.S. Treasury would be honored to the amount of $200,000, would, I was assured by Messrs Fraser, Trenholm & Co, be of no value in a commercial transaction. They expressed themselves disposed, however, to do everything for me in their power."[2]

★ ★ ★

Huse visited the London Armoury Company to purchase Enfield rifles. When his initial purchase of rifles fell short of the actual need, Huse reported to Gorgas:

"I made enquiries at the London Armoury Company for Enfield rifles to be manufactured by them. This establishment is in some respects superior to every other musket manufactory in the world, and in every aspect is equal to the government works at Enfield. It seems to me highly important to obtain rifles from this company if possible. I found that they were willing to entertain a proposition for 10,000, but not anything less than that number.

"After conferring freely with the commissioners and receiving from them an entire approval of my action, I proposed to take from the London Armoury Company 10,000 Enfield rifles of the latest government pattern, with bayonet, scabbard, extra nipple, snap cap and stopper complete for £3,16s.6d. This price is somewhat above the limit given in my instructions from Major Gorgas, and I engaged to take 10,000 instead of 8,000.

"Under all the circumstances, I believed myself not only justified, but required to go beyond my orders. The cost of the 10,000 will be about $195,000. I brought with me but $50,000. More has since arrived.

"Even this would have been insufficient for me to do anything had it not been for Mr. Prioleau of the firm Fraser, Trenholm & Co. This gentleman has most generously assumed the responsibility of the entire contract. I beg leave to express the hope that the government of the Confederacy will lose no time in forwarding to me $100,000 to meet payments as they come due."[3]

On June 27, 1861, Anderson accompanied Huse to SIC & Co. where they set up an account and signed contracts. Anderson immediately ordered "2,000 sets of accoutrements."[4]

Huse wrote Gorgas:

"Ready at the same time accoutrements in number nearly equal to the muskets purchased. Everything purchased is of the best quality and has been obtained at low prices as the condition of the market permitted."[5]

Anderson's August diary entries record the purchase of supplies. "Wrote to S. Isaac Campbell & Co. and ordered an additional number of sabers so as to number 1,000."[6] Three weeks later, he wrote, "Bought from Isaacs 10,000 muskets old pattern."[7] The following day's entry read: "Contracted with Isaacs for a lot of (P53) Enfields and for 11,000 English muskets of very good quality."[8]

When Gorgas requested a price estimate on the items Huse planned to purchase, Huse asked for and received an invoice from

SIC & Co. Huse ordered one of each of the items on the invoice and
forwarded them to Richmond to serve as a sample of the patterns he
planned to supply.

Huse's patterns were modeled after British sealed patterns,
which were approved when a sample of the item was deposited with
the appropriate committee, signed, sealed and certified as being the
sealed pattern. Against this sealed pattern, all future purchases would
be compared. Any item showing variations from the norm would be
amended or scrapped.

Huse wrote to Walker:

> *"With the knapsacks is one complete British infantry
> soldier's kit, with the price of each article marked. The
> prices are those at which the contractors are prepared to
> furnish any quantity required."*[9]

At the end of August, Samuel Isaac telegraphed Anderson at
Dover in Kent, informing him that SIC & Co. had secured a contract
with Behrings Bros & Co., (sic) (Barings) who had previously been
supplying the United States government. Barings would supply 4,000
Enfield rifles per month for the next six months. Union agents had
run out of funds and were made to forfeit a security deposit of £5,000
($25,000). This was fortuitous, not only because it served the needs
of the Confederacy, but had the additional advantage of keeping these
arms from being used against them by the invading Union Army.

★ ★ ★

As a gesture of appreciation to Samuel Isaac for all the valuable
service he had provided, Anderson promised Isaac that he would use
his influence with the Confederate government to "obtain(ing) for
him (Isaac) the appointment of Consul General in England in the
event of the success of our cause."[10] Richmond was also delighted by
the involvement of SIC & Co. in the Confederate war effort. Acting
Secretary of War, Attorney General Judah P. Benjamin, wrote to the
Isaacs on March 17, 1862, thanking them:

"Gentlemen: I am in receipt of your favor of January 29th by the Economist, and desire to express to you the deep sense of obligation felt by this Government for the kind and generous confidence you have shown in us when other foreign countries seem to be doubtful, timorous and wavering. You will find, however, that your confidence was not misplaced, and that we have not failed (as far as we could find means) to make remittances to Capt Huse, although not as rapidly as desired; but our difficulties have been great in procuring secure remittances. Enough has been done however, we trust to relieve you from embarrassment or apprehensions. I find from my books that the amounts furnished to Capt Huse have been recently as follows:

"From Jan 20th 1862 to March 7th 1862: $1,261,000.

"Our demands from England will continue to be quite large, and we trust you may find your connection with our young government equally profitable and agreeable."[11]

★ ★ ★

Debts were beginning to accrue, and Huse found his funds depleted. On May 8, 1862, he wrote Benjamin and suggested a new business plan that he had worked out with the Isaacs. SIC & Co. would put the new purchases onto steamers, knowing that the Confederate government or the individual states would purchase the supplies, either in cash or credit.

"S. Isaac Campbell & Co. have allowed my indebtedness to reach the large amount of $500,000, but in no instance have they shown any disposition to delay, on this account, the shipment of goods from the country. On the contrary, when I have desired it, work has gone on both night and day at their establishment, and every desire has been manifested by them to serve the government in every way possible. My funds are exhausted, but I have obtained credit to as large an amount as I am willing to ask for. Messrs S. Isaac

*Campbell & Co., at my request, now undertake to serve
the Confederacy in a slightly different manner. They are
about to forward to the Confederate States war supplies,
all of which have been selected by me and to offer the
same to the government for purchase. In some cases, I
have agreed to a certain price on delivery; in others no
arrangements have been made.*"[12]

By December 1862, Huse and Anderson had purchased
£1,068,722 ($5,343,610) worth of ordnance and quartermaster stores
from British firms with £582,700,7s ($2,913,500) in supplies through
SIC & Co. alone. Added to this was another £117,750 ($588,750) for
rifles from Vienna.[13]

Of this, £613,589 ($3,067,945) had been received by Fraser,
Trenholm & Co., with £572,883 ($2,864,415), still outstanding.

Items shipped to the Confederacy included 129 cannon, 34,731
regulation British accoutrement sets, which consisted of belts (snake-
hook and other), cartridge pouches (cartridge boxes), ball bags and
cap pockets and 34,655 hard frame and other knapsacks with mess
tins and mess tin covers. One hundred thirty-one thousand, one
hundred twenty-nine stands of arms, which included 70,980 P53
long Enfield rifles and 9,715 P-1856 short rifles, frogs and bayonet
scabbards, slings, oil bottles, woolen Confederate battle flags, medical
supplies, farriers and saddlers tools and a multitude of other military
items were also shipped.

On November 1, 1862, SIC & Co. wrote Gorgas:

*"Every exertion has been used on this side to expedite
the transmission of goods. We trust the articles
enumerated will reach you speedily and prove as
serviceable as anticipated. The supplies are the best
description, such as are issued to her Majesty's troops.*"[14]

After Anderson left England aboard the *Fingal*, Huse bought
large amounts of goods not only for the Ordnance Department but
also for the Quartermaster's and Commissary Departments. These
purchases included: 74,006 pairs of boots, 62,025 blankets, 8,675
greatcoats and 170,724 pairs of socks.

At the beginning of December, Gorgas wrote the new Secretary of War, James A. Seddon[*]:

> *"The purchase of ordnance stores in foreign markets on government accounts is made by Maj Caleb Huse, C.S. Artillery, who resides in London, and whose address is No. 38 Clarendon Road, Notting Hill, West London. Major Huse was detailed for this duty in April 1861. His instructions directed his attention chiefly to the purchase of small arms, but his list embraced all the necessary supplies. Under those instructions, he has purchased arms to the number of 157,000 and large quantities of gunpowder, some artillery, infantry equipments, harness, swords, percussion caps, saltpeter, lead, etc.*
>
> *"In addition to ordnance stores, using rare forecast, he has purchased and shipped large supplies of clothing, blankets, cloth and shoes for the Quartermaster's Department without special orders to do so."*[15]

The Isaacs had a large network of suppliers, including the shoe factory leased to Turner Bros, Hyde & Co. Another of the firm's suppliers was Smith, Kemp & Wright of Birmingham, which produced buttons with the SIC & Co. back mark. Other suppliers included:

★ J.E. Barnett & Sons, London (Enfield rifles)
★ Walter H Hindley, London (Cotton sheeting and bagging)
★ Richards & Co., London (Cotton items)
★ Reynolds & Son, London (Wharfage and warehousing)
★ R&W Aston, Birmingham (Enfield rifles and nipple wrenches for the rifles)
★ Fortnum & Mason, London (Food stuffs for blockade runners)
★ Savory & Moore, London (Medicines)
★ John Hall & Sons, London (Muskets and gunpowder)
★ Davenport & Co. (Earthenware and china)
★ Robinson & Fleming (Gunpowder)

---

[*]In 1862, the Confederacy had three secretaries of war. Judah Benjamin was replaced by George Randolph who was replaced by James Seddon.

★ William Essex & Sons (Curriers)
★ John Churchill, London (Medical books) [16]

SIC & Co. opened a branch in Nassau, on New Providence Island in the Bahamas, just to administer the tremendous quantity of goods destined for the Confederacy.[15] Samuel Isaac's eldest son, Henry, and Samuel Isaac's brother-in-law, Benjamin Woolley Hart, were chosen to oversee the new office.

On May 8, 1862, Henry traveled to Richmond for talks with Benjamin. He delivered a letter from Huse:

> *"I beg to be permitted to introduce to you the bearer, Henry Isaac Esq. of London. Mr. Isaac represents the house of S. Isaac Campbell & Co. of London, to which the Confederate government is more indebted for liberal dealing than any other house in Europe, with perhaps the exception of Fraser, Trenholm & Co. of Liverpool.*
>
> *"Messrs SIC & Co. reposing entire confidence in the good faith of the Confederate government at times when there were very few in this country that did, most liberally advanced a large sum of money to enable me to secure a contract for 34,000 Enfield rifles, which rifles, but for their assistance, would have been obtained by the US government."*[17]

At the time of his arrival, Henry was a fine, dashing and outspoken young man of twenty-six. The other half of the team, Hart, had been living in New York City since 1856.

The Confederate shipping agent in Nassau was Louis Heyliger of New Orleans. Heyliger, a successful and accomplished businessman, had been appointed on November 30, 1861, to supervise all transshipments to Confederate ports.

On July 10, 1862, Heyliger wrote to Secretary of War George W. Randolph with some concerns about Hart and Isaac:

> *"I have a word to say in regard to Mr. Hart and Mr. Isaac who represent the interest here of S. Isaac Campbell & Co. of London. They have shown a*

> *disposition to presume on the alleged services they have rendered to the government – services for which, I suppose, they have been mostly amply remunerated. Mr. Hart – for I have had no intercourse with Mr. Isaac – is a resident of New York and, as the uncle of the latter, came out here to superintend the business transactions of the London firm. I do know that Mr. Isaac has openly boasted that the government owes everything to his house, and that his reason for coming to Nassau was to take the business of running the blockade into his hands, and monopolize it to the exclusion of John Fraser & Co., who, as he alleges, has not done one-half for the Confederate government as his concern has accomplished. It may hereafter be a matter of consideration how far the boasted services of SIC & Co. have been offset by the prices they have charged the government for the supplies."*[18]

Nineteen days after Heyliger wrote his letter, Henry Isaac died from yellow fever. This was a third personal loss for Samuel, following the earlier deaths of his first wife and second son.

It was also a big blow for the company, for Henry was to be the driving force in Nassau. Hart took responsibility in accounting for and warehousing supplies shipped from England and the financial correspondence between the islands and Richmond

CHAPTER 10
RUNNING THE BLOCKADE

Huse's first major purchases of arms and equipment were shipped on the *Fingal,* which left Greenock, Scotland, on October 12, 1861. The *Fingal* was purchased in September 1861 for £17,500 ($87,500), and extreme measures were taken to conceal the ship's mission from ever watchful Union agents. She was registered in the name of a British citizen and arrangements were made to load her by a chartered steamer. This kept Confederate supplies off the closely watched railways.

On October 24, 1861, U.S. Secretary of the Navy, Gideon Welles, wrote Flag Officer Louis M. Goldsborough, commander of the blockading squadron, urging vigilance:

> *"Authentic information has been received that a large quantity of rifles, powder, swords and munitions of various kinds were shipped near London on board the steamer Colletis, which left the Thames on the 29th ultimo for Greenock, Scotland, when her cargo was to be transferred to the new iron-screw steamer Fingal. The transshipment is made to throw this government off its guard. The Fingal is schooner rigged, with two masts, has a round stern, the bust of a man for a head, has one deck and a poop, is 186 feet long and 25 feet wide, and 12-9/10 feet depth of hold. She is British built and her tonnage is a little short of 500 tons. Her cargo consists of 31,000 pounds powder, 525,000 cartridges, 1,550,000 percussion caps, 1,500 rifled "Brown Bessies," 300 sword bayonets, a large quantity of paper for cartridges and other articles much needed in the states of insurrection. A contract has also been made in England for a larger amount of similar articles to be shipped by another vessel, which was to follow within two weeks of the Fingal. The Fingal will, of course, sail*

under an English flag and will undoubtedly attempt to enter one of the Southern ports."[1]

Welles had learned of the *Fingal's* cargo and whereabouts through a network of Union agents and detectives based in Britain. Ambassador Charles Francis Adams was the spy's ring leader. He was also instrumental in maintaining British neutrality and in thwarting diplomatic recognition of the Confederacy.

Anderson recorded in his diary:

"October 10, 1861. We have received information from a reliable authority that it became absolutely necessary to drop the Fingal down the (Mersey) river some miles to avoid the espionage and interference of Mr. Adams' spies. Many of our packages had to be left behind, and the last shipment of 2,000 muskets sent on from London had not been taken aboard. We had been advised by our friend in the Foreign Office that the American minister (Adams) *had obtained an order to send an official to Glasgow fully empowered to ascertain the ownership of the vessel and cargo, and we had anticipated his arrival there by hurrying Mr. John Low* (Assistant to Confederate Naval purchasing agent in England, Captain Bulloch) *with instructions to Mr. Byrne to transfer the ship, cargo and all to Mr. Low – for one shilling and, when called upon by the gov't agent, to announce the fact that the venture belonged to Mr. Low."*[2]

The mock sale worked and saved the *Fingal* and her cargo. She safely sailed on October 12, 1861, and was loaded with some 11,000 Enfield rifles, 500,000 cartridges, over 1,000,000 percussion caps, 3,000 cavalry sabers, 500 revolvers, 2 Blakely cannon, 8,000 shells, 400 barrels of gunpowder, 9,982 yards of blankets, drugs, plus 2,000 rifles for the states of Georgia and Louisiana. Added to this were accoutrements furnished by SIC & Co. and over £10,000 ($50,000) worth of accoutrements and leather from Alexander Ross & Co. Also on board was Anderson, having completed his mission. The *Fingal* arrived in Savannah on November 14, 1861.[3]

Another shipment was placed aboard the steamer *Gladiator*. After her arrival in Nassau in December 1861, her weight and subsequent slow speed caused her to be trapped in the harbor. A Union gunboat – tipped off to the *Gladiator's* arrival – blockaded the harbor and any attempt to make a run for it became impossible. The *Gladiator's* cargo was transshipped to the lighter, faster steamers, *Kate* and *Cecile*, which made for the unguarded port of New Smyrna, Florida.

Heyliger wrote Benjamin:

> *"The steamer Kate arrived here on the 18th and, according to your instructions, I immediately arranged for the transshipment of the Gladiator's cargo, and, as previously intimated, I had no difficulty in obtaining the requisite permission from the authorities. I superintended the transshipment of the following portion of the Gladiator's cargo now on board the Kate, as per the bill of lading enclosed."*[4]

The *Kate* was loaded with: 300 cases of Enfield rifles, (6,000 rifles), 32 bales of blankets and serge, 94 boxes of mess tins, knapsacks and pouches (Enfield cartridge boxes), 514 boxes of Enfield cartridges and 90 cases of percussion caps.[5]

The *Kate* arrived safely in New Smyrna on January 31, 1862. The *Cecile* arrived with the remainder of the cargo on March 2, 1862.

Another of Huse's purchases was shipped aboard the steamers *Economist* and *Southwick*. The *Economist* departed England in February 1862 and sailed first to Nassau before traveling to Charleston, South Carolina. She arrived on March 14, 1862.

SIC & Co. invoices dated December 14, 1861, for "Blue Gray Army Cloth,"[6] December 16, 1861, for "Oxford Gray Army Cloth,"[7] and invoices dated December 23 and December 31, 1861, for accoutrements attest to the size of the shipment.[8]

On February 14, 1861, the United States Minister Resident to Belgium reported that "the cargo of the *Economist* is looked upon as one of the most important yet dispatched."[9]

F.H. Morse, U.S. Consulate in London agreed:

> *"The Economist and Southwick have near 40,000 Enfield rifles, with a large quantity of powder, rifled cannon, army clothing, etc. on board for the Insurgent States."*[10]

SIC & Co. chartered the steamers: *Southwick, Gladiator, Sea Queen, Sir William Peel, Springbok, Stephen Hart* and *Harriet Pinckney* to ship their Confederate orders.

Three of these vessels, the *Southwick, Stephen Hart* and *Harriet Pinckney*, were owned by SIC & Co. Proof of ownership comes in a letter written by Prioleau to Captain J.H. North:

> *"I have received your private letter of the 5th instant and that of your firm, the latter formally turning over to me the vessel lately built by Messrs Fawcett, Preston & Co. In reply to your request that I should take charge of the ship referred to, I have to say that I accept the vessel and request that you will cause a crew to be shipped for her, capable of navigating her across the Atlantic, and that you will have the ship sent to the port of London. This is to be done in case she can be put under English colors and her name to be changed to the Harriet Pinckney. In taking out the register for the ship, her owners may be given as S. Isaac Campbell & Co., who are prepared to sign such papers as may be necessary to make the transfer legal. SIC & Co., as you are aware, are (sic) the owners of the S (Southwick) and they are willing to charter or purchase the Harriet Pinckney."*[11]

On January 29, 1862, bad news came with the capture of the *Stephen Hart* by the Union gunboat *Supply* off the Florida coast. She was carrying a full cargo of accoutrements, cloth and armaments. The proof that she was bound for the Confederacy was so overwhelming that the Isaacs' attorney did not defend their ownership rights.

On March 19, 1862, the *Stephan Hart's* cargo was examined by appraiser, Orison Blunt.[12] The cargo consisted of: 1,546 yards of gray army cloth, 11,543 yards of steel mixed gray cloth for uniforms, 625 gross CSA buttons, 2,220 water proof covers for mess tins, 4,000 ball bags and belts, 15,432 pairs of stockings, 592 pairs of russet blucher shoes, 762 pairs of black blucher shoes, 1,750 white blankets, 6,800 gray blankets, small arms and ordnance supplies.

After the capture of the *Stephen Hart*, Huse began to worry about the safety of Nassau as a viable port. He wrote Gorgas:

> *"The Port of Nassau has become so dangerous even as a port destination for arms in British ships, that I have thought it prudent not to order anything more to that port, for the present at least. I have accordingly ordered the Harriet Pinckney, with arms, artillery and carriage on board to Bermuda."*[13]

Gorgas was in full support of the base. He wrote Huse:

> *"Bermuda will answer very well as a depot. With two steamers now running between it and Confederate ports, if a third can be added, we will be able to bring stores away rapidly."*[14]

Huse appointed Mr. S.G. Porter as the Confederate agent in Bermuda. Porter had been recommended to Huse by Bulloch.

In late 1862, the streamer, *Justitia*, sailed with the largest shipment of the war and arrived safely in Bermuda in December.

The original SIC & Co. invoice spanned 36 pages and included the follow items:

- ★ Surgical instruments, telegraph wire, brandy, 20 sets of scales
- ★ Blankets, cloth, facing cloth, greatcoats and buttons
- ★ 30 waxed leather backs for ball bags
- ★ 300 leather pouch middlings
- ★ 300 leather belt middlings
- ★ 250 leather waist belt middlings
- ★ 500 leather knapsack sling middlings
- ★ 642 pounds cut leather for knapsack trimmings
- ★ 150 leather middlings for gun slings[15]

In mid-1862, SIC & Co. attempted to enter into discussions with the Confederate government for the supply of vast amounts of clothing and equipment. Hart wrote to Randolph:

"My friends, Messrs S. Isaac Campbell & Co. of London, instructed me to address your proposals to furnish clothing and equipment for 100,000 men, deliverable at Bermuda or Nassau, which have failed to reach you owing to the capture of the vessel by which they were forwarded. I now take the opportunity to repeat the proposals to the following effect;

Messrs S. Isaac Campbell & Co. request me to state that there exists now greater facilities for procuring the materials required by the Confederate government than prevailed in the early part of the war, when they were scarce and difficult to procure. Under these circumstances, they feel confident of their ability to give satisfaction to the Department in respect to the quality of the equipment and the dispatch which they can be completed, which they would be able to accomplish in about three months from receipt of order."[16]

John B, Jones, a clerk working in the War Department, noted in his diary that "...A Mr. Hart, agent for S. Isaac Campbell & Co. London, proposes to clothe and equip 100,000 men for us and to receive certificates for specific amounts of cotton. This same house has, it is said, advanced as much as $ 2,000,000 on our account. This looks cheering. We have credit abroad."[17]

In Richmond, Quartermaster General Abraham C. Myers expressed his concerns:

"Respectfully returned to the Secretary of War

"I have considered this communication and submit that it contains nothing upon which to base a report. The proposal is to clothe 100,000 men, but no item of cost of the clothing is presented. Major JB Ferguson of the Quartermaster's Department is now in Europe for the purpose of making purchases for the Department. The parties can make their proposals directly to him, and he can best determine upon the expediency of accepting the proposition of these parties."[18]

★ ★ ★

The *Stephen Hart* turned out not to be the only ship lost. On February 3, 1863, while making for the harbor of Nassau, (despite Huse's warnings) the *Springbok* was captured by the Union gunboat *Sonoma*. The cargo that was captured was significant:

☆ 220 bags of coffee and 300 chests of tea
☆ 4 cases of ginger and 19 bags pimento
☆ 10 bags of cloves and 60 bags of pepper
☆ 53 packages of medicines and saltpeter
☆ 18 bales of army blankets, butternut color
☆ 7 bales of army cloth, 20 bales gray cloth
☆ 4 bales men's colored travelling shirts
☆ 540 pairs gray army blankets and 24 pairs white blankets
☆ 360 gross brass navy buttons, marked CSN
☆ 10 gross army buttons marked C
☆ 397 gross buttons marked I
☆ 148 gross buttons marked A

All the buttons were marked on the underside, S. Isaacs Campbell & Co/71 Jermyn St. London.

The cargo also included: 8 cavalry sabers, 11 sword bayonets, 992 pairs of army boots, 97 pairs of russet brogans and 47 pairs of cavalry boots.[19]

The letter of instruction to the ship's captain read:

"London, December 8, 1862
"Captain James May

"Dear Sir – Your vessel being now loaded, you will proceed at once to the port of Nassau, NP, and on arrival report yourself to Mr. B.W. Hart there, who will give you orders as to the delivery of your cargo and any further information you require.
"We are, dear sir, etc. Speyer & Haywood."[20]

✯ ✯ ✯

What caused one blockade runner to be caught while another made multiple voyages without ever seeing a Union warship? The laws of probability certainly played a role, as did the craftiness of the ship captain. Under the command of Lt. John Wilkinson, the famous Confederate blockade runner, *Robert E. Lee*, made twenty-one successful runs through the Union blockade. One of Wilkinson's tricks was to fly the Stars and Stripes when on the open sea. He once exchanged salutes with the captain of a Union gunboat that passed closely by.

The *Robert E. Lee* was painted to blend in with the ocean, smoke stacks were angled back and the engines burned anthracite coal (which produced less smoke), unless a smoke screen was to their advantage. Wilkinson once had his engine crew shovel in coal dust to create a thick cloud of smoke around dusk; while the Union ship chasing him steered toward the smoke, he altered course and slipped away into the night.

Wilkinson only attempted to enter the harbor on moonless nights. One sailor complained about the danger of colliding with another ship while running at night without lights.

Wilkinson replied, "Perhaps, but if we run our lights, we will collide with a cannon ball."[21] They proceeded without lights and with a good lookout posted.

On November 9, 1863, The *Robert E. Lee* was captured by the Union Navy and re-named the USS *Fort Donelson*.

✯ ✯ ✯

How profitable were blockade runners? A good example of the lucrative (though potentially short lived) nature of the blockade running trade was the ship *Banshee*, a British owned vessel which operated out of Nassau and Bermuda. She was captured on her seventh run into Wilmington, North Carolina. However, at the time of her capture, she had already turned a 700% profit for her owners, who quickly commissioned and built the *Banshee No. 2*, which soon joined the firm's fleet of blockade runners.

One well known toast went as follows:

"Here's to the Southern planters who grow the cotton; to the Yankees that maintain the blockade and keep the prices of cotton up; and to the Limeys who buy the cotton. So, three cheers for a long continuance of the war and success to the blockade runners!"[22]

It was believed that with as little as two successful runs through the Union blockade, a ship had already made a profit for its owners.

The *Robert E. Lee*
(Courtesy of The Civil War Navy Sesquicentennial)

CHAPTER 11
THE ERLANGER LOAN

By late 1862, SIC & Co. was feeling the financial pinch. Hart wrote to Gorgas and enclosed a statement for unpaid cargoes and munitions.

> "Abstract of statement showing the amount due to S. Isaac Campbell & Co., contractors for CS Ordnance stores, dated November 1, 1862, forwarded by Benjamin W. Hart, Agent of S. Isaac Campbell & Co., Nassau; enclosed in a letter of Colonel J. Gorgas to Secretary of War James A. Seddon, December 26, 1862, Richmond Va.

"August 23, 1862 balance, £45,038 ($225,190)
"July 15, 1862 – Steamship Columbia, £44,500 ($222,500)
"August 27, 1862 – Steamship Harriet Pinckney, £120,162.7 ($1,250,813.5)
"August 27, 1862 – (Steamship) Ouachita, £8,000 ($40,000)
"October 1, 1862 – Rifles, etc., Vienna, £175,000 ($875,000)
"October 1, 1862 – Scabbards, etc., £12,000 ($60,000)
"October 30, 1862 – Justitia shipment, - £120,000 ($600,000)
"November 1, 1862 - Steamship Cornubia, - £24,000 ($120,000)
"November 1, 1862 - Steamship Justitia, - £20,000 ($100,000)
"November 1, 1862, - Cornubia Shipment, - £8,000 ($40,000)
"Interest - £6,000 ($30,000)

TOTAL: £582,700,7s ($2,913,500)"[1]

The company had laid out over £400,000 ($2,000,000) and was pushing for settlement of at least part of the £582,000 ($2,910,000) bill Huse and Anderson had run up. On November 17, 1862, Hart wrote to Randolph: (Unbeknownst to Hart, Randolph had resigned from office that very day and had been replaced by Seddon.)

"*Messrs S. Isaac Campbell & Co. having up to this time advanced to the government about £400,000 sterling, their active means are so far diminished as to require that some portion of the payment be made in cash or sterling bills of exchange.*

"*Following my instructions, I beg leave at the same time to bring to your attention the state of the account between Messrs S. Isaac Campbell & Co. and the government for equipments and other war material already furnished. They have been given to understand the government has matured plans for the issue of certificates for certain specified quantities of cotton, which are to be made applicable to the liquidation of debts contracted abroad for war material and for future purchases.*"[2]

On December 5th, Gorgas wrote Seddon:

"*To pay for these purchases, funds have been from time to time sent to him (Huse) by the Treasury Department, on requisitions from the War Dept, amounting in the aggregate to $3,095,139.18. These have been wholly inadequate to his wants and have fallen far short of our requisitions. He was consequently in debt at latest advices to the amount of £444,850, a sum equivalent, when the value of exchange is considered, to $5,925,402 of our currency. While this capacity for running in debt is the best evidence of the ability of Major Huse, the debt is a matter that calls for immediate attention. It is respectfully urged that immediate steps should be taken by the Treasury Dept to supply funds for the liquidation of this indebtedness.*"[3]

With no major funds forthcoming from the Treasury Department, Hart took his case to Secretary of State Benjamin.

"*I am instructed by our London house most respectfully but most earnestly to pray your attention to* *our account for military supplies furnished to your government, which has now reached the amount of half a million sterling and will soon extend to £600,000 ($3,000,000), when orders now in hand for ordnances are executed. We feel assured that you will so well understand the urgent necessities for money pressing upon our house under such circumstances that so far from needing any apology for asking large remittances on account of it at the very earliest moment, you will most cordially concur in acknowledging our claims to the attentive and prompt consideration of the cabinet and the government.*

"Since the very incipiency of the struggle in which the Confederate States have been and continue to be engaged, we have afforded them every assistance in our power, and we refer with unmixed satisfaction to the testimonials and kind acknowledgements from yourself when administering the War Department of the services we have rendered to the state in the infancy of the war, when our resources in money and credit were placed without limit at its disposal and probably contributed in some degree to the success of its armies in the field.

"We sincerely and earnestly trust that the claims of our house are not forgotten or, if temporarily in oblivion, will be revived by our present and earnest appeal to the attention of the government. During the months that have passed since the date of your letter, the indebtedness of the government to us has been largely increasing, and we have continued to furnish supplies and to hope on from month to month that the remittances which we were promised would arrive.

"In conclusion, we again most earnestly crave your attention and that of the cabinet and government to our solicitations, at the same time embracing the opportunity to assure you that we are desirous of continuing to place our services and our resources at the disposal of the government of the Confederate States."[4]

In another letter to Benjamin written on the same day, Hart stressed the importance of prompt payment by the Confederate government:

> *"Referring to our previous letter of this date, we feel ourselves impelled still further to represent to you the extreme urgencies of our house for money and the critical condition in which we are placed from the absence of the remittances we have so long and so anxiously looked forward to, and it is not without great pain we apprise you that we find ourselves much embarrassed, which may be increased to the most serious point unless we receive the most immediate relief from the government.*
>
> *"We venture further, most respectfully, to represent to you that a crisis in the affairs of our house would, in our humble opinion as mercantile men, exercise a most pernicious and withering influence on the credit of the Confederate government in foreign countries for the abandonment of friends whom you have so kindly mentioned as 'having exhibited toward the government a kind and generous confidence at a moment when all others in foreign countries seem to be doubtful, timorous and wavering' could do no otherwise than produce an impression not easily to be effaced from the sensitive minds of commercial and moneyed men.*
>
> *"We venture also to hope that our representations may receive the attentions of the cabinet, and that immediate relief may be dispatched to us. Our immediate wants are £120,000 ($600,000), and we trust to receive a considerable portion of this by return of steamer, the balance of our account in the succeeding month at furthest, and at least the same amount in the same periods ensuing. The receipt of these funds will enable us to proceed with our business with advantage and place us in a position to enable us to continue to furnish the government with military supplies."*[5]

The Confederacy had cotton, but now it needed money. With cotton prices rising in England, Richmond decided to use cotton to pay for future purchases. The Confederate government offered warrants for cotton in bulk amounts, and some 400,000 bales were obtained by the government for that purpose. These warrants were priced at eight cents per pound; about a quarter of the price cotton was trading for in London at the time. The first of the new bonds was secretly issued to several European institutions. The proceeds were used by the Confederate Navy for its shipbuilding program in Liverpool.

The idea of secure cotton bonds soon drew the attention of Erlanger & Co., a prominent French investment banking firm that had connections all over Europe.

Rumors of the arrangement had been circulating around Europe since early 1863 and had made their way back to the United States through diplomatic channels. The U.S. Ambassador to France wrote to Secretary of State William H. Seward:

> "A correspondent of the London Post says he learned that a Confederate loan for five millions £ sterling has been negotiated through the house of Erlanger and Co. in conjunction with the leading capitalists of London and Liverpool...the truth of this statement is at least doubtful. Your means of judging it are as good or better than my own."[6]

A month later and better informed, the Ambassador wrote Seward that he was "...inclined to believe that arrangements have been made with the House of Erlanger & Co., Frankfurt, to loan the Confederates three millions £."[7]

★ ★ ★

Who was Erlanger & Co.? According to the U.S. Ambassador to France, the House of Erlanger was a German Jew house.[8] The house was more than a Jew house. Baron Emile Erlanger was one of the wealthiest men in Europe, and this German (and Jewish) banker had branches in Amsterdam, Paris and Frankfurt.

The cotton bonds came to Erlanger in December 1862 after the Confederacy failed to find a British bank to float them. After several weeks of negotiations, the financier sent three agents to Richmond to propose a much larger bond issuance, as cotton was then selling for sixty to eighty cents per pound. The bonds were to be redeemed for cotton at the rate of ten cents per pound. The Confederates agreed, and the Erlanger or Cotton Loan as it became known was able to raise the sum of $15,000,000 or £3 million for the Confederate government.

The bonds were denominated in £ Sterling and French francs and offered in the European cities of Amsterdam, Frankfurt, Liverpool, London and Paris. The exchange rate was to be $5.00 (Confederate) = £1 Sterling. These were actually very favorable terms as in 1860 the exchange rate was $4.86 (US) = £1 Sterling. The terms would never be that generous again. There are historians who speculate that if Richmond had followed Erlanger's advice to leverage the cotton bonds, the Confederacy would have had a permanent history.[9]

The bonds sold at 90% face value and were redeemable for government controlled cotton in the South. This provision stimulated blockade running because the holders of Erlanger bonds had to risk the blockade to convert the bonds into cotton.

When the subscription opened on March 19, 1863, the bonds enjoyed success, with the most prominent and wealthy people in England making investments. A brief list of British subscribers included:

* S. Isaac Campbell & Co., of 71 Jermyn Street, London, army contractors, £150,000 ($750,000)
* D. Forbes Campbell, 45 Dover St, Piccadilly, London, £30,000 (150,000)
* Alex Collie & partners, £20,000 ($100,000)
* Sir Henry de Houghton, Bart., £180,000 ($900,000)
* Thomas Stirling Begbie, 50 Mansion House Place, London, £140,000 ($700,000)
* The Marquis of Bath, £50,000 ($250,000)

A full list of subscribers can be found in Appendix E.

Altogether, subscribers bought £1,759,894 ($8,799,470) in bonds. The subscription sales might not have been so robust but for an incorrect assumption on the part of British investors. Confederate agents encouraged the rumors that the Erlanger bonds would be honored no matter how the war ended. Thomas Dudley, the U.S. Consul in Liverpool, was appalled at both the brazenness of the Confederate agents and the naivety of local merchants. He reported to Seward:

> "...as strange as it may seem, these people here who are aiding the Rebels and taken or purchased these bonds believe if the worse comes and the Union is restored, the United States government will assume payment of their bonds."[10]

The U.S. Consulate allowed that the Erlanger Loan had an immediate and positive effect on Confederate purchasing operations. In fairness to the investors, who badly miscalculated the munificence of Uncle Sam, the Confederate loans were still backed by $45 million worth of cotton.

On May 25, 1863, Emile Erlanger, Jr., on behalf of Erlanger & Co., made an agreement with SIC & Co. to take over the company's account and recover the money owed (£515,000 or $2,575.000) them by the Confederate government.

As part of the agreement, Huse would give a certificate of indebtedness to SIC & Co. In return, Erlanger would give SIC & Co. £150,000 ($750,000) in bonds for which they had already subscribed and another £150,000 in bonds, plus a £90,000 ($450,000) cash advance.

SIC & Co. was to leave in the hands of Erlanger & Co. the cotton warrants and eight percent bonds now deposited with the latter for the security of the advance. SIC & Co. was to obtain a full £300,000 ($1,500,000) of Erlanger bonds, with half of the bonds to be delivered before June 15, 1863. It was agreed that SIC & Co. would remain responsible to the Confederate government for their account, and that SIC & Co. was to have their account audited by the Confederate's Chief Financial Agent, Colin McRae. SIC & Co. would not dump the bonds onto the market while the investigation into their affairs was taking place.

On the following day, Huse wrote to Erlanger & Co.:

"Gentlemen:
"S. Isaac Campbell & Co. have an account with the Confederate States government for army supplies furnished by them on my order, upon which account there is due a balance of, say, £515,000 ($2,575,000). This amount I agree on the part of the Confederate States government may be transferred to yourselves upon any terms that may be agreed upon between yourselves and Isaac Campbell & Co., and I further agree that, provided the account is thus transferred to you, all the money that I may receive from any source for the payment of this claim shall be paid to you."[11]

SIC & Co. presented their sizeable bill to Erlanger & Co. and made the following agreement for payments of sums due.

"Copy of agreement between Messrs. S. Isaac Campbell & Co. and Messrs. Emile Erlanger & Co.

"London, May 25, 1863
"We, the undersigned, Isaac Campbell & Co., of London, and Emile Erlanger & Co., of Paris, have made today, the following agreement on behalf of the claim Isaac Campbell & Co. state to have against the Confederate government for the sum of about £515,000 ($2,575,000). Major Huse, who incurred this debt with Isaac Campbell & Co., will give them a certificate of indebtedness for the above sum, which they will hand over to Emile Erlanger & Co. Emile Erlanger & Co. engage to hand over to Isaac Campbell & Co. £150,000 ($750,000) of paid-up bonds for which they have subscribed, and for which they will credit Emile Erlanger & Co. with £135,000 ($675,000), less the discount allowed to subscribers, this quantity being considered paid up.
"Isaac Campbell & Co. will receive, furthermore, from Emile Erlanger & Co. £150,000 ($750,000) in bonds, the

former crediting the latter with £135,000 ($675,000), no discount being allowed to these; £90,000 ($450,000), which Emile Erlanger & Co. have advanced to Isaac Campbell & Co., £90,000 ($450,000), together with interest, will be taken over from their account and credited to Emile Erlanger & Co. Emile Erlanger & Co. will pay in cash £40,000 ($200,000) to Isaac Campbell & Co., and on the 2nd of June, the further sum of £20,000 ($100,000); and Isaac Campbell & Co. will have the right to draw on Emile Erlanger & Co. at ninety days date for the sum of £40,000 ($200,000), and place this to the credit of the latter, less interest for seventy days.

"Emile Erlanger & Co. engage to deliver the above-mentioned £300,000 ($1,500,000) of bonds as soon as they are ready, and half of these no later than the 15th of June. Isaac Campbell & Co. will leave in the hands of Emile Erlanger & Co., to be returned to the government, the cotton warrants and eight percent, bonds now deposited with the latter for the security of their advance of £90,000 ($450,000).

"In the event of Isaac Campbell & Co. wishing to resell these bonds, they shall give the refusal of the bonds to Emile Erlanger & Co. at such price offered by other parties and which Isaac Campbell & Co. are willing to accept.

"Isaac Campbell & Co. remain responsible to the government for their account, and hold Emile Erlanger & Co. free from any loss or reclamation in this respect; and they furthermore leave the balance of their account as a guarantee and will not claim it from the government for two months. Isaac Campbell & Co. will allow their account to be audited by General McRae or any person he may appoint on his behalf.

"Emile D'Erlanger
"For Emile Erlanger & Co.
"S. Isaac Campbell & Co."[12]

Huse wrote to McRae:

> "As security for the payment of their account, I have deposited with them cotton warrants representing £100,000 ($500,000) and eight percent bonds for $2,000,000. Upon these securities Messrs Erlanger & Co. have advanced to S. Isaac Campbell & Co. the sum of £90,000 ($450,000). Messrs Erlanger & Co. have further arranged with S. Isaac Campbell & Co. to take over their entire account. Please reimburse Messrs Erlanger such amounts as may be due to them on the final settlement of their account. The cotton warrants and eight percent dollar bonds they will return to S. Isaac Campbell & Co., by whom they will be given up to me. These securities and all money that I may receive from any source on account of the Confederate States government I will turn over to you, to be held until the amount of payments made for me under this arrangement shall be covered by proceeds from sales of cotton and treasury drafts drawn in my favour. I further suggest not to make any new purchases on government accounts until these claims are satisfied, except with your previously obtained approval."[13]

CHAPTER 12

FERGUSON AND CRENSHAW

By the middle of 1862, it had become apparent to the Confederate War Department that additional purchasing agents were needed for the different departments – especially the Quartermaster's Department. This idea for a separate purchasing agent came from Assistant Quartermaster General Larkin Smith, who wrote Randolph:

> "To obtain a full supply of clothing for the army is becoming more embarrassing and difficult as the raw material is diminishing and the machinery employed in the manufacture becomes worn out. Every exertion has been made to render all resources of the country available; but if, in the matter of clothing and shoes, there were ample supplies of the raw material, the capacity to manufacture them is wanting, thereby, rendering it certain that reliance upon our own sources of supply will be in vain. In view of these facts, I respectfully recommend as the only effective mode of relief from these difficulties and embarrassments that an officer be dispatched to Europe to purchase cloth, shoes, blankets and other indispensable articles of issue to the troops.
>
> "The government would save largely by purchasing abroad, even if one of every three cargoes were lost. To depend upon private enterprise to import these goods is to trust a very unreliable source of supply and pay enormous profits to the importers.
>
> "Messrs Ferguson, of this department, who has been employed in providing materials for the clothing department at Richmond, is fully competent to purchase goods abroad, as in addition to his knowledge of quality and prices, he has the mercantile ability and integrity to disburse advantageously the large sums which would be

entrusted to such an agent as may be sent to Europe. The necessity of almost immediate arrangements for these supplies leads me to ask your early consideration on this subject."[1]

The man selected by Smith for this task was sixty year old Virginian Major James Boswell Ferguson, Jr. Ferguson benefited from more than twenty years experience in the mercantile profession, having his own import and export business, Ferguson JB, Jr. Bros & Co. Cloths, Cassimeres and Vestings. He was also a man of stature having married Emma Cabell Henry, the granddaughter of Patrick Henry.

Ferguson – carrying a bureau draft for £533,000 ($2,665,000) in sterling and $1 million in Confederate bonds – was posted to Manchester in October 1862 by Myers. Ferguson and Myers knew each other well. Ferguson had been Myers' principle contracting agent in Richmond and was responsible for securing quartermaster goods from Southern mills before his departure to England.

Ferguson arrived in England in late December and based his operations in Liverpool. He transferred his base to Manchester to be nearer the woolen mills of Lancashire and Yorkshire. His first orders were for 1,000,000 yards of woolens, with half in blue for trousers, 500,000 yards for greatcoats, plus 300,000 blankets, 150,000 wool hats and 600,000 pairs of socks.[2]

Ferguson, quickly found himself in direct conflict with Huse, who refused to give up the responsibility of not only buying supplies for the Ordnance Department but also for the Quartermaster's Department, and all of it through SIC & Co. Due to Huse's mounting debts, Gorgas secured an order from the War Department directing Ferguson to stop purchasing and to turn over his funds to Huse until Huse's debts were paid off. Ferguson did as ordered.

Ferguson wrote Lawton, who, in August 1863, had become the Confederacy's second Quartermaster General. Ferguson complained that: "Huse has caused me more annoyance than all the others combined on this side and has defeated my plans in a great measure for keeping up the supplies of our department."[3]

Ferguson convened a meeting with Huse to instruct him that as the Chief Agent for the Quartermaster's Department in Europe, he would resume responsibility for purchasing for that department. He examined some of the items Huse had purchased and became concerned by what he saw. He wrote to Myers citing irregularities in Huse's dealings with SIC & Co.

"In the first interview I had with Major Huse, he informed me that his indebtedness to Messrs S. Isaac Campbell & Co. was, in round numbers £500,000 ($2,500,000) the true figure by this time was £515,000 ($2,575,000) and that over £100,000 ($500,000) had been used for the QM Dept. He (Huse) said it was true that upon some of his purchases, he had received a commission, but he intended to use a part of the money to pay his travelling expenses, and the balance, amounting to £1,000 ($5,000) to purchase a military library, which he intended to present to the Ordnance Dept. I expressed the opinion that the amount of commission should have been deducted from the face of his invoices in the shape of a discount, and that I would advise him to postpone his donation to the Ordnance Department until his debts were paid and our army shod and clad.

"After the meeting was over, a conversation occurred between an officer of the navy and myself, the sum and substance of which you will find in a copy of his letter enclosed. The next morning I called at Major Huse's office, and he showed me the invoices of the articles sent out by the Justitia. My familiarity with some of the classes of goods mentioned in said invoice led me to believe that extortionate prices had been charged for them.

"I requested Major Huse to show me a sample of the 12,000 yards sent out at 7s.6d per yard. I took a sample of it and feel no hesitation in saying that a similar article can be furnished at from 4s.6d to 4s.10d per yard, equal in every respect to the cloth sent out. From the foregoing, you will perceive there were three prominent facts for me to consider; firstly the admission of Major Huse that he had received a commission on some of his purchases; second that the senior partner of the house

*through whom nearly all of his business had been
transacted offered to divide a commission with an
officer of the navy, and third that exorbitant prices had
been charged for such articles as I could identify."*[4]

The naval officer was Lieutenant James H. North. Secretary of
the Navy, Stephen Mallory, had ordered North to England in May
1861 to assist Bulloch in purchasing ships for the Navy. North
recounted the incident to Ferguson:

*"Dear Sir: Your letter of March 30th has just been
received. In that letter you ask me to do you the favor to
state in writing the substance of a conversation I had
with you shortly after your (my) arrival in this country,
touching on an offer made me by S. Isaac, of the firm S.
Isaac Campbell & Co. to divide a commission with me on
a business transaction for the government of the
Confederate States, and whether or not I regarded that
offer as an attempt to induce me to combine with him (S.
Isaac) for the purpose of defrauding the Confederate
States government, and whether I rejected the same on
that ground. In reply to the foregoing, I would say that
the subject of the conversation to which your letter refers
may be briefly stated as follows:*
*"I did call on S. Isaac of the firm S. Isaac Campbell &
Co. on a matter of business; that Mr. Isaac did in the
course of conversation make an offer to divide with me a
commission of five percent on a business transaction
with the Confederate government, and that I did regard
that offer as an attempt to induce me to enter into a
transaction to defraud the Confederate government, and
that I did reject the offer."*[5]

These commissions were in fact the way business was
traditionally conducted by purchasing agents on large contracts with
British commission houses. For example, on September 27, 1861,
Major Edward Anderson was offered a commission by Alexander
Ross & Co. In closing a deal for purchases for some £10,000
($50,000), Anderson wrote:

"Whilst I sat conversing with him, Mr. Ross quietly passed over to me a cheque for two hundred and fifty odd pounds, payable to myself individually, as a return commission for my transaction with him... I knew very well that this was the English way of doing business and that the government permitted its officers to receive these commissions."[6]

Anderson accepted the check, but signed it over to the Confederacy in the form of a credit note. For his part, North refused the offer, but Huse accepted the commissions offered by SIC & Co.

★ ★ ★

At about the same time Ferguson left for England, Seddon agreed to a proposal by Frank Ruffin, who worked in the Confederate Bureau of Subsistence, to form a partnership with William G. Crenshaw of Richmond to establish a line of private blockade runners. Crenshaw, along with his brothers, owned one of the largest textile mills in Virginia.

Per the terms of the agreement, the War Department acquired cotton to export in the name of the Crenshaw brothers, while Crenshaw would contract for European ships. The agreement between Crenshaw and the War Department also gave Crenshaw and his brothers one quarter interest in the ships, with the government holding the balance.[7]

Even though a joint venture between a profit motivated company and the War Department was not ideal, Seddon believed the plan was the best way to get supplies to the Confederate Armies. He wrote of Crenshaw, "I am satisfied of the character and energy of Mr. Crenshaw and am only solicitous that the faith of the Department be observed with him."[8]

Crenshaw arrived in England in January 1863 and entered into a partnership with Alexander Collie – a Scottish shipping agent – who had offices in both Manchester and London. Crenshaw proposed an equal partnership with Collie in the building of a steamship line. The other half of the company would be owned by the War Department. Collie advanced the funds for the first vessel.

Half the ship's cargo space was designated for the War Department and one quarter reserved for the Navy. That left a quarter of the space free to carry whatever luxury goods Crenshaw and Collie determined. The luxury goods would be duty-free, while the government items promised a commission of 2-1/2%.[9] Each run through the blockade would generate enough profit to buy an additional steamer. Crenshaw gained Ferguson's support for the venture.

Seddon chose to turn a blind eye to the obvious profiteering of Crenshaw and Collie believing that any ship bringing in supplies was better than none. He wrote Crenshaw:

> *"I prefer you should continue to give your experience, energy and judgment to the conduct of a business in which, I hope, with reasonable profits to yourself and your associates, you will be enabled to render valuable service to the cause of your country."*[10]

All of this was done without Huse's knowledge and brought Ferguson, Crenshaw and Huse in competition for supplies. Huse still operated as though he had carte blanche to purchase items for all departments. Therefore, he refused to sanction Crenshaw's purchases in order to keep the business with SIC & Co. He also believed that any sizeable cargo space not devoted to military stores was not aiding the Confederate war effort. He wrote Crenshaw:

> *"Dear Sir,*
> *"Referring to the conversation I had with you on the subject of your mission to Europe, I have to say that in compliance with the instructions of the War Department, I will keep you informed of the wants of the War Department as I may from time to time receive them...   "As regards to the purchase of supplies for the Ordnance and Medical Departments, I shall make the purchases without availing myself of the services of Messrs Crenshaw & Collie, excepting in such cases as I may feel satisfied their agency would be advantageous to the Confederate States government."*[11]

Chapter Twelve

Three days later, Huse wrote Crenshaw again:

> "In the purchase of army supplies....much better can be transacted by the house with which I have had my large transactions, which transactions have received the unqualified approval of the War Department"
>
> "...In communication with you on Saturday last, I informed you that I was not prepared to place the purchasing of the Ordnance and medical supplies in your hands... I have not received any instruction from the War Department from which I can draw the influence that I am to do so.... The government has already four steamers: the Giraffe, Cornubia, Merrimac and Eugenie engaged in running the blockade, and I have instructions to purchase a fifth. These steamers would have to be idle or be sold if I were to turn over to you the purchasing and forwarding of supplies for the Ordnance and Medical Departments."[12]

Frustrated by Huse's lack of cooperation, Crenshaw wrote directly to Seddon:

> "Since my meeting with Major Huse I have been very much embarrassed by the course I ought to pursue. Every officer here to whom I have had occasion to explain the arrangements I am endeavouring to carry out expresses himself as highly pleased with it except Major Huse. Then naturally I am led to inquire why does it not meet his views? I have no hesitation in saying that it is because it takes from the hands of Isaac Campbell & Co. the purchase of the government goods.
>
> "This is the true, and, in my opinion, the only reason. Why is he so anxious to retain this business in their hands? He says because they have been so liberal with our government. I say, no; it is not in their nature to be liberal. They have never had credit here for anything of the sort, and when it is told they have advanced £500,000 ($2,500,000) for 2-1/2 percent commission it bears its own falsity in its own face. They were formerly

contractors with the English government, but were dismissed as such, and their contracts cancelled by the Secretary of War in May 1858 for alleged bribery of one of its officers. They remonstrated and tried to explain it was a loan of £500 and not a gift to the receiving officer, but the Secretary of War adhered to his determination and refused to reinstate them.

"It is true that they went before a committee appointed to examine into the corruption of the Crimean War generally, and in 1859, on the evidence of one of the firm (S Isaac), the committee reported it was a loan to the officer, although there was no evidence taken of the debt and was altogether a very loose transaction. As far as I can learn, the English government has since ventured to do but little with them directly.

"You have doubtless before you evidence that they offered to bribe one of our own officers last year. You know whether there was anything in Captain North's character to justify them in making to him such a proposition without daring to make the same to others who had been dealing largely with them for more than a year.

"At all events, the scorn with which Captain North refused it showed that they would have been quite as safe to have made such a proposal to any one else. Major Huse admitted both to Major Ferguson and Captain North that he had received some commission on government purchases since he had been here, which he intended to apply to the payment of his expenses here and the purchase of a library to send as a present to the Ordnance Bureau, but finding the government so pressed for money, he had paid the amount over to the credit of his account with Isaac Campbell & Co.

"Mr. White, a commissioner sent here by the state of North Carolina, who has had some opportunity of seeing something of Isaac Campbell & Co., informs me that he entertains of them the same opinion that I do; nor have I seen any man since my arrival here who would say a good word for them except Major Huse.

> "Believing it absolutely necessary that all government business here should be under one control, when I heard of the appointment of Major Huse, I thought it a move in the right direction and met him (without prejudice, except that he was from the North) with every desire to co-operate with him.
>
> "I am satisfied from what I see and hear that he is not fit for the position, and I sincerely trust it will be your pleasure to select someone now in the Confederate States of high character for integrity and honor, of great business capacity, to come over here and take charge entirely of the financial and commercial affairs of our government. Let the orders of every description come directly to him, and by him executed through that party that he thinks will do it best.
>
> "He should have entire control of the finances here, with discretionary power to apply the funds (when enough for all purposes are not to be had) to those in his opinion the most important. Of course it is difficult to find such a man, but we have found a man fit to be president of the Confederate States and others to form a cabinet, we can find a man fit to occupy the position.
>
> "I suggest that we can find him, too, among our native born Southerners.
>
> "I remain my dear sir, yours ever truly,
>
> "William. G Crenshaw" [13]

Seddon, receiving letters from both Crenshaw and Ferguson alleging the payment of commissions by SIC & Co. to both Huse and North, decided to investigate the matter. He wrote Gorgas:

> "This matter appears of serious nature. The taking of a commission is altogether inconsistent with the purpose and duty of a trusted agent of the Department. The matter should be fully investigated."[14]

Huse wrote Gorgas:

> "I have not only been personally annoyed by the conduct of Major Ferguson and Mr. Crenshaw, but my efficiency as an agent for the CS War Department has been seriously impaired to such an extent that I think it is important that my character for integrity and soundness of judgment should be fully re-established; or failing this, that some other officer should be detailed for the important duty to which I have been assigned." [15]

Gorgas was stunned by the charges leveled against Huse. He wrote Seddon:

> "Major Huse is an officer of nearly fifteen years service. He knows perfectly well that the naked transaction of taking a commission on purchase, on receiving, directly or indirectly, compensation for purchases for the government would dismiss him from the service with disgrace; yet he makes confession of this flagrant crime to a stranger in his very first interview with him. It is unnecessary to suggest the propriety of at least hearing Major H.'s statement..."
>
> "The matter of Major Huse's unfitness for making purchases is assumed by the Quartermaster General probably on the testimony of Major Ferguson. I think it proper to say that I am perfectly satisfied with his business capacities and so far as that is concerned, desire no change." [16]

Gorgas ordered Huse to confine his purchasing to ordnance and medical supplies until the dispute was sorted out. He also ordered Huse to settle outstanding debts with the firm:

> "There was a distinct agreement that if £125,000 ($575,000) was paid, the balance would be taken in cotton on this side. The debt of S. Isaac Campbell & Co. being settled, permit me to suggest to you the propriety of closing with this house and making your purchases as

far as possible with cash without the intervention for commission; or, if that is unavoidable, to select some other house."[17]

When SIC & Co. refused Gorgas' terms, he wrote Huse:

"The house of SIC & Co. must be held to their agreement made by their house in Nassau. The agreement was that on receiving £125,000 ($575,000) in cash, the remainder of the debt due them would be paid in cotton. The money has been paid according to that agreement, viz; £75,000 ($375,000) in sterling, sent to you, and £50,000 ($250,000) in gold. Receipted by the attorney of SIC & Co., Mr. Battersby at Savannah. It is probable that the prices charged by SIC & Co. were made with due regard to risks incurred on our account for articles purchased. This was not only inevitable and just, but was to us an invaluable resource."[18]

Gorgas acknowledged the risks the company had taken and the support SIC & Co. had given the Confederacy, but he reiterated his warning to Huse.

"No one house should monopolize the business of government or any branch of it, but that purchases should be made of cash, directly of parties who offer the most advantage."[19]

Under pressure from Seddon, Gorgas appointed Colin McRae to examine the "accounts and vouchers of Huse,"[20] and ordered McRae to pay attention to the financial transactions of SIC & Co.

★ ★ ★

Seddon failed to grasp the obvious conflict of interest with businessmen and profiteers in the mix with Confederate purchasing agents or the potential all this had for interdepartmental rivalries of the sort demonstrated here. He was primarily interested in getting the supplies from Europe to the Confederacy, even if he had to ship with privateers to achieve this goal. He continued to ship with privately owned blockade runners, as well as on the three or four government owned vessels. As a result, the Confederacy would struggle to meet all of its needs until McRae took over the shipping of supplies in 1864.

CHAPTER 13
THE DISPATCHING OF COLIN MCRAE
AND THE DELAYED ARRIVAL OF
M. HILDRETH BLOODGOOD

On May 26, 1863, the day after SIC & Co. made their arrangements with Erlanger & Co., Colin McRae was officially appointed to investigate Huse's business practices.

McRae was born in Sneedsboro, North Carolina, on October 22, 1813. His family moved to Mississippi, and, after his father's death in 1835, McRae took over the family's banking and finance companies. After serving a term in the Mississippi legislature, he moved to Mobile, Alabama, and established Boykin & McRae, (later Boykin, McRae & Foster) a cotton trading company. At the outbreak of the war, McRae worked on the Mobile defenses and helped to establish the Selma Arsenal. He was an original member of the Confederate Congress and signed the Constitution of the Provisional Government of the Confederate States.

In July 1862, McRae became an agent in the Confederate Ordnance Bureau. He was sent to Europe to act as the Chief Financial Agent for the Erlanger Loan. His first task was to create a system of credit so the Confederacy could continue purchasing needed supplies.

Following his appointment, McRae was instructed by Gorgas to pay particular attention to the contracts Huse had made with SIC & Co., since this was where the allegations of impropriety resided. McRae announced his intentions to Huse:

> "I am in receipt of a communication from Col J Gorgas, Chief of Ordnance, informing me that I have been appointed to examine your accounts as the disbursing officer of the War Department abroad. Please let me know when and where you will have your accounts ready to place before me. At any time after the

The Dispatching of Colin McRae and the Delayed Arrival of M. Hildreth Bloodgood

1st of August, I shall be ready to commence the investigation."[1]

Since overseeing the Erlanger Loan was a full time job and the impending investigation into Huse's affairs was a distraction. McRae received word from Seddon that M. Hildreth Bloodgood had been appointed to assist him in the investigation. Seddon ordered McRae to wait for Bloodgood's arrival before proceeding with an examination of Huse's accounts and SIC & Co.'s books.

★ ★ ★

Like Huse and Gorgas, Matthias Hildreth Bloodgood was a Northerner. He was born in New York in 1825. He successfully ran the Mobile office of the family business. In 1853, he married Augusta Kennedy, a Mobile native. Three of their four children died – two from diphtheria within a week of each other. On November 20, 1862, Augusta died. She was twenty-nine.

After Augusta's death, Bloodgood continued to live in the family home. In the summer of 1863, he faced conscription into the Confederate Army. To avoid being drafted, he wrote to a friend in Richmond asking him if he could "obtain some minimul (sic) employment which would enable me to go to Europe."[2] News reached Seddon, who decided to employ Bloodgood's expertise to aid McRae in the investigation of Huse's dealings with SIC & Co.

On June 1, 1863, five days after McRae's appointment, Bloodgood was ordered to England. After the war, Bloodgood would write of this appointment: "My friend sent me instructions from J.A Seddon, to aid C.J McRae in the examination of the accounts of the main disbursing officer in Europe."[3]

Seddon wrote to Bloodgood confirming his appointment and giving him his initial orders:

"...The Department hereby authorizes you to examine the accounts of Major Caleb Huse of the Ordnance Department. The appointment is made subject to the following instruction: Some short time ago, the Chief of Ordnance applied on behalf of Major Huse the

appointment of a person for the examination of his vouchers and accounts. I suggested the name of General CJ McRae as a suitable person for the purpose, this was acceded to and has been communicated to him. It is supposed that Gen McRae may be otherwise employed and may not be able to give all the attention necessary.

"Therefore, the Department has concluded to approach you in the hope, and in the event that General McRae cannot act or is unable to do so, you are hereby authorized to do so alone... The Department prefers that General McRae and yourself should act together and you will, therefore, communicate with him on your arrival."[4]

It would take at least a month for Bloodgood to arrive in London. But Seddon had been clear. The audit could not proceed until Bloodgood arrived. Concerned about the delay, Huse wrote McRae:

"A letter received from Col Gorgas by me yesterday on the subject of my accounts appears to be more full than the one which you received by the same conveyance, and which is referred to in your note of yesterday. I therefore inclose (sic) a copy of it. Mr. Bloodgood has not arrived, and it being uncertain when he may be expected, and really important, as I conceive, that as early an examination as possible should be made, I beg to suggest that you request some other gentleman from the Confederate States to discharge the duty expected by the War Department of Mr. Bloodgood.

"I am only desirous that the examination should be made as soon as practicable, and in view of the effect that the charges brought against me by Major Ferguson in his letter to you may have in lessening the confidence of the businessman in me until these charges are found to be untrue, it seems to me highly important for the interest of the government that the examination should be made as promptly and as thoroughly as possible.

"As matters now stand, important business negotiations may at any time be interrupted by

interested parties circulating the story that I am an officer under charges of malfeasance. An attempt of this kind has already been made, as you are aware, by William G. Crenshaw. I beg, therefore, both as a matter of public importance and private interest to myself, that you will take steps for having at least a preliminary examination made at an early day."⁵

The next day, McRae wrote Gorgas to explain the latest situation:

"On the morning of the 22nd instant, I received a communication from you dated the 26th of May, informing me that I had been appointed to examine the accounts and vouchers of Maj. Caleb Huse, the special agent and disbursing officer of the War Department abroad. On the receipt of this communication, I addressed a note to Major Huse informing him that I would be ready to commence this examination immediately after the 1st of August and asked him to have his accounts ready to lay before me at that time.

"In the afternoon of the same day, Major Huse called on me and bought with him another communication from you from which the Secretary of War had appointed M. Hildreth Bloodgood to make the examination in connection with me. I thought I would await the arrival of Mr. Bloodgood before commencing the examination.

"As it is uncertain when Mr. Bloodgood will be here, I have concluded to proceed with the examination as I proposed. It is due to Major Huse that the investigation should be made as soon as practicable, besides the public interests require it, as his usefulness is impaired by charges that have been made against him.

"In making the examination of the accounts and vouchers of Major Huse, I shall ask the assistance of H. O. Brewer, Esq., now in Paris, a gentleman of high character and a thorough businessman, familiar with all sorts of accounts and devoted to our cause."⁶

McRae and Brewer commenced their examination of Huse's vouchers and accounts. The investigation was constantly delayed because McRae was called away to deal with other government financial business. He conceded as much to Gorgas on August 28, 1863:

> *"Every facility for the examination has been afforded by Major Huse and Messrs S. Isaac Campbell & Co., but owing to the very nature of the transactions and the difficulty of obtaining confidential clerks, we have not thus far called in any other assistance, consequently our progress has been rather slow. We have just finished the examination of the accounts current and find them based on correct business principles, accurately made out, and sustained by the proper vouchers; but we have not yet gone into the business of prices, which will be a tedious and laborious business, and much care will be required to arrive at a proper decision, as the prices of most of the articles were much higher during the period of purchase than they were before or have been since, owing to the great competition in the English markets for all the military supplies which then existed."*[7]

On August 28th, Bloodgood arrived in Southampton from Havana. He made his way north to the Leamington Spa instead of to London to link up with McRae and Brewer. From Leamington, Bloodgood wrote McRae: "I arrived yesterday per steamer (from Havanna) (sic) at Southampton. I enclose you a letter which will explain itself. As I understand it, it is for you to decide whether my services are needed and if so what shape they are."[8] McRae responded and asked Bloodgood when he would be ready to commence the examination.

McRae also wrote Gorgas: "If I do not hear from him (Bloodgood) by Monday, the 6th, when I will be prepared to go on with the examination of the accounts, I will address him another note asking his assistance."[9]

On September 15th, McRae, after rendezvousing with Bloodgood, updated Gorgas on his lack of progress in the investigation.

*"You should be advised of what has been done on this
side of the water in reference to the settlement of the
account of Major Caleb Huse with Messrs S. Isaac
Campbell & Co. I think that the settlement of this account
was very favourable to the government, as it enabled us
to dispose of £300,000 ($1,500,000) of the stock of the
(Erlanger) loan at the issue price. But I fear that it will be
disastrous to Messrs S. Isaac Campbell & Co., as they
have the whole of this stock, representing £267,224,15s.
10d ($1,336,220) of their account on hand; also £50,000
($250,000) in gold in the Confederacy, amounting to an
aggregate to £317,224,15s.10d, ($1,586,120) none of
which is available unless they were to force a sale of
stock.*

*"This would entail a loss of £80,000 or
$400,000...Great as this sacrifice would be, Mr. Saul
Isaac (the financial partner of the house) informs me
that unless they can get early relief from our
government, they will be compelled to make it, as it will
be impossible for them to meet their engagements with
so large an amount locked up in Confederate securities.*

*"The result is that this house, which has been so much
maligned by our overzealous friends, is likely to be
ruined by having trusted our government when nobody
else would... I have seen Mr. Bloodgood and arranged
with him to take up the examination of Major Huse's
accounts on the 17th (Sept) instant... We shall continue to
avail ourselves of the valuable assistance of Mr. Brewer
and hope by the end of two weeks to make a final report
on the subject."*[10]

The immediate effect of the delays was to undermine Huse's
ability to do his job. Gorgas was stuck in the middle between the
Secretary of War, the Quartermaster's Department (Ferguson and
Crenshaw) and Huse.

CHAPTER 14
THE INVESTIGATION

James A. Seddon
(Courtesy of the Library of Congress)

By mid-October 1863, the examination into Huse's accounts and vouchers was complete, and the investigation into the charges brought by Ferguson and Crenshaw finally got under way. McRae invited Ferguson to present his evidence:

"Sir,

"In June last, you addressed me a letter making charges against Major Caleb Huse. I replied, referring you to the War Department and enclosed a copy of your letter to Richmond. I have since then, at the Secretary of War's request, undertaken (with the aid of M.H. Bloodgood) the examination of Major Huse's accounts.

"We will be in London for some days and would be glad to hear from you personally if possible, if not by letter with full details of your charges. The former course would be preferable, as having the papers before *us, we could confer more understandingly. We would ask your immediate attention and aid, as our stay here is but limited, and the Department desires, for many reasons, a thorough examination and a speedy report."*[1]

Ferguson sent the following replies to McRae and Bloodgood:

"Answers to questions forwarded by General C.J. McRae:

"Question 1: "What was the amount of the commission and from who received?

"In answer to the foregoing, I shall give the substance of a conversation with Major Huse shortly after my arrival. He said on some of his purchases he had received a commission. I explained my surprise that he should have done so, and I expressed the opinion that the amount of commission should have been deducted from the face of the invoice in the shape of a discount. He said that it was the way of some of the brokers to divide with parties giving them orders as part of their commission. He also stated that he intended to pay his travelling expenses and the balance amounting of £1000 ($5,000) or so he intended to use in purchasing a military library to be given to the Ordnance Department. I ask that the evidence of the Hon. J. Mason and Capt. North of the Navy Dept be taken on this point.

"Question 2: What evidence have you of this specification? What goods were then changed in the invoice and what proof have you on the fact?

"In reference to the first specification to my charge No. 2 suspecting goods sent by the Justitia charged at 40% or 50 % above the market price of the day. I submit the following. In looking over the invoice of the above mentioned goods, my attention was particularly struck with an item of 12,000 yards of blue grey army cloth

85

charged at 7/6 per yard at a commission of 2-1/2%. I took a sample from a piece of cloth which Maj. Huse informed me was the same kind as sent out by the Justitia. I submitted the sample to several highly respectable manufacturers. I asked them at what price they would make a similar cloth.

"I now submit copies of their written tenders, also various other samples of my own purchases with prices attached, which I think will prove to you the truth in charge of this specification.

"After stating the foregoing, I asked to be furnished with the invoice of the Justitia as handed to me by Mr. Bloodgood is not a correct copy of the invoice exhibited to me by Maj Huse last December, the commission is left out... In further proof of which, I beg you will swear Mr. Thomas Bayne, who was present with me at the office of S. Isaac Campbell & Co. where this invoice was exhibited. I ask that Mr. W. Crenshaw and Mr. Charles Hobson be sworn on the subject of commission.

"Your obedient servant,
"Maj. J.B. Ferguson."[2]

Due to the complexities of the case and amount of transactions, McRae and Bloodgood found the going difficult. In a letter to Seddon, McRae admitted his struggles in uncovering the truth.

"Since my last of September 15th to Col. Gorgas, Mr. Bloodgood and myself have spent much time upon these accounts and yet have not been able to make an examination complete enough to report to you our final conclusions. We are both here in London and have made arrangements with an accounting and examining house which I think will enable us to arrive at nearly certain conclusions.

"Major Huse has shown from the very first every desire to aid us in every way in his power. My object is simply to report the fact that we are at work and making as much progress as the enormous and

complicated nature of the transactions admit, and will, at as early a day as practicable, send you an official report."3

The London accounting firm McRae appointed was Quilter, Ball & Co., headed by William Quilter and John Ball. The firm was the top auditing and accounting firm in London at the time and had participated in the investigation of the Weedon Bec scandal in 1859.

After the examination of his own accounts and vouchers had been completed, Huse submitted SIC & Co.'s books to McRae and Bloodgood. SIC & Co. had already agreed to have their books audited in return for receiving payment from the Erlanger Loan. In the following correspondence to McRae and Bloodgood, Huse explained:

"The charges against me appear to be limited to my transactions with the house of S. Isaac Campbell & Co. It has been assumed that no one could be sufficiently well acquainted with the miscellaneous articles embraced in my purchases of and through that house to make the purchases understandingly, and that I have therefore placed myself completely in the power of that house, perhaps innocently, but at all events to the prejudice of the interests of the Confederate States government.

"When this investigation first commenced, I foresaw great difficulty in the way of making it thorough, which I was very anxious it should be, and expressed my anxiety to Messrs S. Isaac Campbell & Co. at the same time that I furnished them a copy of Major Ferguson's letter to Mr. McRae denouncing me as a dishonest agent of the government.

"They appreciated the difficulties of the case and expressed themselves desirous of doing anything in their power to enable me to vindicate my character. They have offered to exhibit their books to the auditors appointed by the Confederate States government to examine my accounts and to give them the means of tracing every transaction they have had with me from the date of my first order, not only for goods purchased

*by them on my order, but for everything sold to me from
their own establishment.*

*"My collection of samples has been made entirely for
my own guidance. Not expecting such an examination as
this now going on, I find that it will be impossible for me
to provide you with samples of every lot of articles
purchased and to state accurately the price and date of
purchases as you desire. With some articles this can be
done, with others it cannot.*

*"Some articles have been purchased by S. Isaac
Campbell & Co., in small lots as they could get them. In
such cases, I have only preserved samples of lots which
differed considerably either in price or quality, or both.
You will perceive, therefore, that while my collection of
samples is quite sufficient for the purpose intended, and
is indeed serviceable than it would be if the samples
were more numerous, it is not adapted to the purpose
for which you desire to make use of it."*[4]

The broadness of the charges made by Ferguson and Crenshaw
presented difficulty for the auditors in terms of either dismissal or
validation, since the items purchased by Huse over the past two years
were long gone, presumably serving the cause of the Confederacy on
the battlefield. As a result, these goods were not available for the
auditors' inspection as to their quality.

It was Huse who provided McRae with a new direction for the
investigation. In a letter to McRae and Bloodgood, Huse suggested:

*"I beg therefore to suggest to you the only really
efficient means of arriving at the facts you wish to
ascertain, that you will avail yourselves to the first offer
made by S. Isaac Campbell & Co., and carefully examine
all their books. In the course of such an examination, you
would be able to determine the actual profit made by SIC
& Co. on every article, and you could afterward, if you
think proper, continue your examination by applying to
the houses from which Messrs SIC & Co. made their
purchases, the names and addresses of which you could
learn from the invoice book of Messrs SIC & Co."*[5]

McRae, Bloodgood and Quilter, Ball & Co. acted on Huse's advice. From SIC & Co., they collected over 1,000 invoices from suppliers, financial papers and transactions. (These papers and invoice books would be found in McRae's home nearly 150 years later and become known as The McRae Papers.) They also visited SIC & Co. suppliers to ascertain the purchase amount of the goods sold to the commission house.

The more McRae, Bloodgood and Quilter, Ball & Co. delved into the case, the more they were convinced that Huse was innocent, at least in relation to the charges of colluding with the Isaacs to de-fraud the Confederate government. On February 12, 1864, sensing he was nearing the truth, McRae wrote Seddon from Paris where he was overseeing the Erlanger Loan:

> *"With regard to remarks relative to Major Huse, I am hardly ready at this moment to expressing a final opinion, but I deem it due to him to say that although he has made some very serious mistakes, I think there is no good reason to suspect his integrity, and that he has always sought what seemed the best interests of the government, and has, with all his mistakes, really been of great service and done great good, and that you should take into consideration the immense labors which he has been compelled to discharge almost singly, and which forced him to place great confidence in some leading house, which was unfortunately, as you surmise, much misplaced in the case of Isaac Campbell & Co., but I can see no reason to believe there has been any collusion with them."*[6]

In the summer of 1864, after a thorough search of SIC & Co.'s invoice books and an examination of the accounts of all the various suppliers to SIC & Co., McRae finally received the evidence he needed to prove that fraud had been committed by SIC & Co. The commission house had two sets of books. One invoice book showed the actual cost of the items bought, while the other invoice book contained the prices quoted to Huse. Not only had the Isaacs charged more than the agreed upon 2-1/2% commission, they also overcharged Huse between 5% and 20% on the majority of the items bought.

The Isaacs explained the overcharges by claiming that the Confederate government had not paid their debts promptly. They also claimed that cargoes had been lost to the blockade. Not only that, but the firm had advanced large sums of money for purchases and had been waiting patiently for payment. As a result, the extra charges were nothing more than an insurance policy against the risks involved.

McRae demanded a refund of the overcharges. The Isaacs refused. McRae gave the invoices and account books to Thomas & Hollams, the law firm that represented the Confederacy in England. Meetings were held with the Isaacs, but a settlement was not forthcoming. In an August letter, Thomas & Hollams informed McRae that the best course of action would be to take the case to arbitration.

"Dear Sirs,

"Isaac Campbell & Co. v Confederate States:

> *"Referring to our interview of this morning, we do not gather that the (illegible) initiated by Maj Huse and Messrs Isaac Campbell & Co. was intended as anything more than the basis of a settlement and consequently that it cannot be relied upon as a concluded agreement putting an end to all questions. Assuming this to be the only alternative seems to be to proceed with the arbitration. It is to that our opinion remains unchanged, namely that it is practically useless and it would be unwise to attempt to limit the claims of Messrs Isaac Campbell & Co. and that of the representatives of the Confederate States of America to agree to an arbitration.*
>
> *"The only course is to leave Messrs Isaac Campbell & Co. unfettered as to the demands which they make and to trust to the good sense and discretion of the arbitration as to be made in which any attempts which they may make to depart from accounts and claims*

already ventured. Of course, the Confederate government is in no way legally bound to submit any matter to arbitration, nor can they be sued in this country.

"Consequently, they have it in their power, if they choose to avail themselves of their position, wholly to defeat any attempt on the part of Messrs Isaac Campbell & Co. to enforce their demands as Messrs Isaac Campbell & Co. are perhaps not unreasonably pressing for a definite reply.

"We shall be glad to receive instructions whether or not we shall proceed with the proposed arbitration and prepare the draft of agreement of reference.

"Yours Faithfully,
"Thomas & Hollams
"Messrs Quilter, Ball & Co."[7]

★ ★ ★

In his final report to Seddon, McRae admitted Huse had made mistakes when he placed all his business in the hands of one commission house, but he also praised Huse for his efforts in purchasing the supplies that kept the Confederate Armies clothed during the opening years of the war. He also suggested Huse receive a pay raise.

On October 1, 1864, McRae and Bloodgood forwarded their official report to Seddon:

"...we understand from Major Huse that he has received from home an official communication endorsing his conduct. It was not our intention in that dispatch to clear him from blame, but to relieve him from any charge of collusion and to place before you the difficulties he had to encounter, and the good he has done and tried to do as palliative or offsets against his errors and mistakes. We wished to keep from speaking too severely of his mistakes because of the difficulties of his position, but not to endorse or overlook them."[8]

Although arbitration between SIC & Co. and the Confederate government continued until the end of 1864, SIC & Co.'s direct dealings with the Confederacy ended. A comparison between the invoices from 1863 and the previous two years reveal that only a trickle of supplies were shipped from SIC & Co. after McRae's arrival.

SIC & CO.'S 1863 EXPORTS

April 11 – For Medical Dept – Medical and Medicine books

May 12 – Bought of W&W Webster £266 ($1,330) worth of books

October 22 – 16 bales of Humnals (sic)

October 23 – 1 piece of blue gray cloth 58 ½ yards

October 27 – 3,000 Enfield Rifles, 750 nipple wrenches and 150 bullet molds.

November 12 – 30 bales lined tarpaulin

The only major purchase was on June 30th for 28,389 Austrian rifles, 28,389 bullet molds and 28,400 scabbards for a total cost of £100,284,17s.6d. ($501,420)[9]

However, the Isaacs continued to sell goods to the Confederacy. On November 29, 1864, Turner Bros sold 3,000 pairs of army shoes to Major J.F. Minter, Chief Quartermaster of the Trans-Mississippi Department. On December 19, 1864, he purchased another 1,400 pairs, the same day Ferguson purchased fourteen bales of gray wool shirts.[10]

Chapter 15
S. Isaac Campbell & Company
The End

When the Confederacy collapsed in May 1865, SIC & Co. still held £300,000 ($1,500,000) in Erlanger bonds[1]. The bonds were now worthless. The Federal government would not back them or release the cotton for liquidation by the creditors. The Isaacs, along with a significant number of other European investors in the Erlanger bonds, were *raptus regaliter*.[2]

The arbitration case concerning the Huse overcharges never reached settlement before the Confederate government was officially defunct.

SIC & Co. hired Speyer & Haywood to bring suit in the United States in an effort to recover losses suffered when the *Stephen Hart* and the *Gertrude* were captured attempting to run the blockade. The *Stephen Hart* was ostensibly headed for Cuba, via Nassau. The ship was captured off the coast of Florida on January 29, 1862. This was the last time a vessel using only wind for locomotion would try to run the blockade – it was too easily caught. The *Gertrude*, a British blockade runner built in Scotland, was captured near Eleuthera Island as she headed for Charleston in April 1863. Both were condemned as lawful prizes and contraband of war.

The case was heard in the Southern District Court of New York. SIC & Co.'s attorneys argued that the vessels were neutral property and ought to be returned to the rightful owners. They further argued that the cargo was not intended to be received by belligerents or foreign agents. Hart was not in Nassau as agents of the Confederate government, but to conduct the shipping business of the firm. The Court ruled for the United States as the Plaintiffs had not come to the bench with clean hands based on their past activities.[3] Coming to the bench without clean hands can be used as an affirmative defense in a court of equity and is a bar to recovery and proved so here.

The Court's decision read in part: "it is satisfactorily established that the cargoes of both the *Stephen Hart* and *Gertrude*, were, when captured, on their way to the enemy's country, into which they were designed to be introduced by a breach of the blockade, and as S. Isaac Campbell & Company was interested in the entire cargo... this inference is regarded as a very proper one and warranted by the proofs invoked."[4]

The Isaacs appealed the case to the U.S. Supreme Court, and the case was heard on January 11, 1866.

SIC & Co.'s attorneys argued that the ship was neutral, flying the Union Jack, and Britain was a neutral government. In addition, it was a commercial schooner bound for a neutral destination – Cardenas, Cuba. An additional contention was that the Union blockade did not extend to Cuba. All this really meant (in fact) was that the *Stephen Hart* had not yet reached port, and the cargo on board had not been conveyed on to a speedier vessel, as was the case with the *Gertrude*.

On March 26, 1866, the Court published its decision, ruling that neutrals who engage in belligerent trade with contraband cargo under the cover of a false destination cannot complain if their ships are seized and condemned as enemy property.[5]

SIC & Co. also brought suit to reclaim the *Springbok*, which was boarded on February 23, 1863, approximately 150 miles off the coast of Nassau. U.S. sailors found the ship under charter by Thomas S. Begbie and carrying a cargo belonging to SIC & Co. (both familiar names to the sailors), but the vessel carried no cargo manifest and, based on the unsatisfactory condition of the papers, the vessel was seized.

Unlike the *Stephen Hart* or the *Gertrude*, the *Springbok* carried mostly non-contraband, with only a small part of its cargo consisting of arms and ammunition. The owner was May & Co., and the captain was James May, a son of the owners.

T.S. Begbie's charter called for a "voyage to Nassau with a cargo of lawful merchandise goods, the freight to be paid one half in advance on clearance and the remainder cash on delivery; thirty running days to be allowed for loading at the port... and discharging at Nassau."[6]

The bills of lading did not indicate who the owners were, nor did they disclose the contents of more than a third of the packages. The manifest also failed to reveal the nature of the cargo, and both manifest and bills of lading consigned the cargo to order. No invoices were found on board. A letter from Speyer & Haywood to Captain

May instructed the latter to report to Hart on arrival at Nassau. Hart would give him orders as to the delivery of the cargo and any further information required.

In the examination *in preparatorio*,[7] Captain May stated that he did not know the reason for the *Springbok's* capture, neither was he aware that there were any contraband goods on board. The Prize Court ruled that his testimony was a study of ignorance.

The Court decided that the *Springbok* was subject to seizure. "When the same claimants intervene for different vessels or goods... the prize courts inquire, did they come engaged in traffic similar to that which they are charged in that particular case."[8] In other words, what was the claimants' intent?

In 1863, however, after the decree of condemnation had been issued, but before the judgment of the U.S. Prize Court was delivered, the Isaacs initiated an appeal through the British Government. The Law Offices of the Crown had opined that the sentence was unjustifiable both to ship and cargo, but indicated that they did not intend to do anything about it. Lord Richard Lyons, British envoy to the United States, wrote:

> *"Her Majesty's Government was therefore disposed to think that the sentence was wrong and ought to be reversed, but they would nevertheless be glad to see the reasons upon which it was founded, and they would hesitate to instruct your Lordship to interfere in the case until it had been heard before the Court of Appeal... For the above reasons, therefore, and upon a full consideration of the whole case, Her Majesty's Government do not feel that they would be justified, on the materials before them, in making any claim on the United States government for compensation or damages on behalf of the owners of the cargo of the Springbok."[9]*

In 1866, the US Supreme Court proved more liberal than the Prize Court and restored the *Springbok*. Chief Justice Chase applied the Court's former ruling in the seizure of the *Bermuda* with respect to the cargo, which was worth £66,000 ($330,000).

"...where decision of goods destined ultimately for a belligerent port are being conveyed between two neutral ports by a neutral ship, under a charter made in good faith for that voyage, and without any fraudulent connection on the part of her owners with the ulterior destination of the goods, the ship, though liable to seizure in order to [effect] the confiscation of the goods, is not liable to condemnation as prize."[10]

★ ★ ★

While trying their separate cases in the U.S. Courts, SIC & Co. continued to do business supplying the British Volunteer units and whatever small arms deals around the world they could attract. Overall, the business of military arms brokerage was no longer as robust as during the war and would never be again.

Burdened by the huge losses from the Erlanger bonds, lacking any great success recovering funds from the U.S. Courts and unable to re-establish any business with the British Army, SIC & Co. filed for bankruptcy in 1869. A deed dated June 21, 1869, declared:

"The inspectorship under Thomas & Hollams.

"Samuel and Saul Isaac, both of No. 7 East India Avenue, Leadenhall Street, City of London, merchants and co-partners trading under the style or firm of S. Isaac Campbell & Co. Two names of non-creditors are listed: Edward Barnett of 134 Minories, London, gun-maker; William Palmer of Bermondsey, Surrey, leather merchant. Their role implied 'whereby the joint and separate estates of the debtors are respectively agreed to be wound up and distributed under the inspection of the said Edward Barnett and William Palmer in all respects of bankruptcy.'"[11]

At the age of sixty-eight, Samuel Isaac acquired the rights from the promoters of the Mersey Railway Tunnel connecting Liverpool with the Wirral Peninsula. He raised the funds to complete the tunnel to great fanfare, and when he passed away on November 22, 1886, he left an estate valued at £203,084,17s.9d ($1,015,420).

Saul Isaac became a respected mine owner. His involvement with the colliery business can be found in two articles from the *Jewish Chronicle*. The first article was published on June 10, 1870.

> *"The new colliery at Clifton, sunk by Sir Robert Clifton, also absorbed a lot of capital. Coal was proved in 1867, and the first sod was cut on 19th June 1868. Sir Robert died shortly after the pit was opened. In May 1870, Saul Isaac took a lease on the colliery and, in June, he announced he would meet all accounts which had occurred prior to his taking possession. Fortunately for Mr. Isaac, the boom of 1871-75 coincided with the pit's period of maximum productivity. The works were completed in June 1871 when Isaac announced that 'a good supply of coal can always be had at the pit bank.'"*[12]

The second article was published on July 1, 1870.

> *"A ceremony of great importance to the local interests of Nottinghamshire has just taken place. Mr. Saul Isaac has recently become lessee of the Wilford Bridge and Clifton collieries, with, as the Nottingham Daily Guardian states, the congratulations and goodwill of everybody. Mr. Isaac has inaugurated his proprietorship by two splendid entertainments at Nottingham. It may be mentioned that the first shaft was sunk by the late Sir Robert Clifton, MP, whose widow formally opened the colliery on this occasion."*[13]

The Nottingham Archives show that Saul Isaac purchased land on September 30, 1871, for £1400 ($7,000) from landowner Henry Smith. In 1875, he took a lease for land from the Duke of Newcastle:

> *"Lower Soft and Lower or Bottom Hard Seams of coal in closes called King's Meadows in parish of Standard Hill, Nottingham. For 28 years at £450 ($2,250) per*

annum plus royalty of £75 ($375) per acre for Lower Soft and £120($600) per acre for Bottom Hard coal, plus 1s for every 2,400 lbs. ironstone, but not for unworkable or unmarketable coal."[14]

Saul Isaac (foreground) outside his Colliery in Nottingham.
(Courtesy of the Nottingham Central Library)

His mines provided much needed employment for the region around Nottingham. Over 4,000 people signed a petition asking him to run as the Conservative Party candidate for the borough of Nottingham in the next general election. Isaac agreed and was elected as a Member of Parliament from January 31, 1874 to March 31, 1880. He was elected to office in the only general election to give Benjamin Disraeli and the Conservative Party an outright parliamentary majority. After his defeat in 1880, Isaac ran again in the 1885 general election in Finsbury Park in London, but narrowly lost.

After the second defeat, his fortunes declined. One newspaper reported that he: "was a handsome man of good presence, but speculative and unstable in his business affairs, and his second period of brilliance proved as fleeting as the first. He fell into difficulties from which he is believed never to have entirely recovered."[15]

The National Archives reveal Saul in and out of court over the years, involved in various lawsuits with banks and business partners.

He died in 1903, considerably less well off than his elder brother Samuel, leaving an estate of just £29.00 ($145).[16]

Chapter 16
Conclusion

The timing of the war proved fortuitous as SIC & Co. needed the business after having lost their military contracts with the British Army. Although the bookkeeping practices of SIC & Co. were questionable, the dichotomy was that the firm remained quite loyal to the Confederacy, even to the point of supplying their own blockade running fleet and advancing their own money when credit was necessary. There is some logic to the Isaacs' claim that the hidden commissions charged the Confederate government were not unreasonable or excessive due to the poor state of Confederate credit.

In addition, there were great risks involved for the firm such as the loss of two SIC & Co. blockade runners to the Union. However, it is hard to get around the fact that if a fair premium for risk was indeed the case, then the full extent of all charges should have been disclosed. The second set of books was damning, for which there was no justifiable explanation.

Looking at the role of Caleb Huse, it is difficult to conclude that he was either naïve or delusional for permitting a firm with the reputation of SIC & Co. to handle most of the CS Ordnance Department purchases without closer supervision. But in Huse's defense, he was a stranger in a foreign land and not acquainted with any of the commercial arms makers or their owners. He needed arms and munitions in a hurry, and the obvious solution was the commission houses which could quickly locate and purchase all he needed.

While Huse's and SIC & Co.'s defense was that splitting commissions was the English way of doing business in the 1860s, it was not ethical. If Huse was not found guilty of any malfeasance in his dealings with SIC & Co., the Confederate government did not intend for him to share commissions from the purchases brokered by SIC & Co., and certainly not to use those funds for his own expenses or a War Department library.[1] In addition, Richmond did not agree to be overcharged for their purchases.

In the final analysis, the Ordnance Department needed military supplies fast and in large quantities. SIC & Co. stepped in to fill that void, and Huse went above the call of duty by purchasing goods for the Quartermaster's Department well before their own agents arrived, even though he was under no obligation to do so. However, by mid-1863, one wonders if Huse had not received enough in commissions for that war library, and why he never mentioned his plans to purchase one to Gorgas.

Some good did come out of McRae's inquiry. It brought to light the helter-skelter purchasing by agents of various departments and commands, which led to competition and resulted in higher prices for the government. In addition, McRae complained to Gorgas that the Confederacy was missing out completely by not undertaking the shipping of supplies. He went as far as to say that, "not a bale of cotton should be allowed to go out of the country nor a pound of merchandise go in, except on government account."[2] Gorgas, for his part agreed, but only after the government had purchased additional vessels.

The point is that additional attention was paid to the efficiency of not only the purchasing operation and the Erlanger Loan, but also to the shipping of goods through the blockade to Confederate ports. And with the improved supervision by McRae, purchases took on the efficiency and organization of the rest of the Ordnance Department. The Isaacs would soon find themselves on the outside and desperate to get their enormous unpaid bills settled.

The quality of some of the goods brokered by SIC & Co. was poor, including cloth of middling quality and accoutrements made from less expensive cuts of leather with marginal craftsmanship. However, there is no arguing that early in the war, without the talent, energy and determination of Huse, and the full financial and logistical backing of SIC & Co., the Confederacy had only limited means to maintain and equip their armies.

One is left wondering in the end, without Huse and SIC & Co., just how long might the newly formed Confederate States of America have lasted, armed with antique muskets, shotguns, pikes and brickbats?

Chapter Sixteen

With McRae taking an active role in the supervision of both ordnance and quartermaster purchasing, only the best quality goods were purchased and at the right prices from foreign sources.

And there was still one manufacturer based in Ireland who was still keen to get in on the act. His name was Peter Tait.

APPENDICES

S. ISAAC CAMPBELL & CO., LONDON

Appendix A

Partial Text of the Report of the Investigation of Weedon Bec Depot

It devolved upon Mr. Elliott, on his arrival at Weedon, on the 7th December, 1855, to organize the establishment, taking the ordnance regulations as his basis and to initiate a system of bookkeeping. We are desirous to keep our account of the system of bookkeeping adopted, as far as possible, distinct from the narrative of the general mode in which the business was conducted; but it is, perhaps, hardly possible entirely to dissever the two subjects. To make either intelligible, it may be convenient to proceed, in the first instance, chronologically with the history of the establishment.

No books whatever had been kept previous to Mr. Elliott's arrival. The only records of the stores previously received at the depot being the bills of delivery from the storekeeper's office at the Tower, Woolwich and other military establishments, in respect of goods sent from those departments, and inspection notes, which accompanied the delivery of goods furnished by contractors. For the accounts which Mr. Elliott had to keep, and the correspondence he had to conduct, he was, for the first five months after his arrival, supplied with only five temporary clerks, "necessarily very young men, with no experience," all perfectly ignorant of the duties of an ordnance station and only one of whom afterwards passed his examination.

Mr. Elliott's own statement is, and we believe it to be true, that seeing the utter impossibility of establishing so large a system of bookkeeping as he would have done with more ample means, the utmost he could do was to subdivide his duties into several branches: the saddlery branch, the boot branch, the cloth branch and the garniture branch, and to direct the foreman in each of these branches to keep an account of the daily receipts and issues. He then "started a ledger as well as he could, in the roughest possible way, and set a clerk to work upon it." The number of clerks was gradually increased.

In July, 1856, there were eight clerks. In November, 1856, they consisted of eleven; and in March, 1857, in consequence of Mr.

Elliott's urgent representation of the necessity of further clerical assistance, three additional temporary clerks were added. These, fourteen in all, constituted the bookkeeping establishment until September 1857 a number wholly inadequate to the rapidly increasing duties which they had to perform.

Besides the rapid dispatch of troops to China and India in the spring and summer of 1857, nearly 50,000 men were added to the army between January 1857 and May 1858, and 30,000 embodied militia was called out in the course of the same year, 1857. This last measure alone doubled in one week the work at Weedon.

On the 1st of September 1857, four clerks were added to the office, and shortly afterwards, Mr. Tatum, an experienced military storekeeper, with an assistant, Mr. Munro, were added to Mr. Elliott's staff. By this time, considerable arrears existed in the books, and upon a representation of Mr. Tatum, strongly backed by Mr. Elliott, of the necessity for further assistance, both in the store and bookkeeping department, six clerks and four persons intended to act as store holders were sent down in October, but none of the latter being conversant with the issue or management of stores, Mr. Elliott appropriated all ten to the bookkeeping department, the duties of which were largely increasing.

According to the Ordnance regulations, the store ledger of every station is made up to the 31st March in each year and should be ready for transmission to the War Office within four months after the expiration of the financial year. It was not considered necessary that Mr. Elliott should transmit the ledger made up to the 31st of March 1856, when he had been less than four months at Weedon. His first ledger, therefore, comprised a period of 16 months ending on the 31st of March 1857 and would have been due at the War Office not later than the 1st of August 1857.

It did not arrive, although he sent up according to regulation a balance sheet purporting to show the amount of stores received, issued and remaining in hand at the station and which should have been compiled from the store ledger. He was called upon for explanation as to this balance sheet and directed to send up the store ledger itself. But he is stated to have "fenced" with this demand, in other words, he urged as an excuse for its non-production, "the deficient clerical assistance" at his command.

In August 1857, it was resolved that Mr. Elliott should be transferred to the post of storekeeper at Dublin and be succeeded at Weedon by Captain Gordon. This change was not resolved on from any suspicion as to Mr. Elliott's honesty; but it was thought that on account of Captain Gordon's military experience, he would carry on the duties in a more satisfactory manner than Mr. Elliott. The change could not, however, conveniently take place until after the Dublin storekeeper had made his annual demand.

Mr. Elliott appears to have continued to make frequent representations of the arrears in his books and the impossibility of overcoming them when the whole time of the clerks was occupied with the current work of the office; and at length, in February 1858, partly in consequence of these representations and partly from the opinion expressed by Captain Gordon of the necessity of changes in the mode of conducting the business at Weedon, Major Marvin was sent down to investigate "the past and present state of the establishment" and, among other things, to inquire into the state of the ledger.

Just before Major Marvin went down, ten additional clerks were appointed and sent down to Weedon, making the total number thirty-seven. But we have been informed by Captain Gordon that, instead of an increase of ten clerks, there should have been an increase of twenty-three: that was the lowest number he agreed with an experienced officer in the War Department in considering necessary; and in February 1858, they made a joint representation to this effect to Captain Caffin, the director of clothing.

Captain Gordon succeeded Mr. Elliott as storekeeper on the 14th of May 1858, when the "remain" or stocktaking of the stores at Weedon was completed, and they were formally handed over to him by Mr. Elliott. The ledger of 1856-7, however, was not quite completed, when, on the 22nd of May 1858, Mr. Elliott absconded, leaving England for America, abandoning his wife and having an actress as the companion of his flight.

The ledger, on its completion early in June 1858, was sent up to the War Office, and after a thorough examination (occupying nearly two months) with the vouchers, 5,400 very few of which were missing, it appeared to be perfectly satisfactory and complete as regards the issues or credit side of the account, while the receipts or debit side of the account only required verification by comparing

them with the accounts of payments made for stores delivered at
Weedon. This comparison and verification has since been
satisfactorily completed. Whether the stores, which according to
these accounts had been delivered at Weedon, were actually received
there, could, however, only be ascertained accurately when the ledger
of 1857-8 up to May 14, 1858, had been completed, and the amounts
then appearing in the ledger to the debit of the storekeeper had been
compared with the actual stores handed over on that day to Captain
Gordon on the completion of the remain.

In order to prepare this second ledger, Commissary-General
Adams, with a staff of eight officers, proceeded to Weedon, but, as
they were not conversant with the system of accounts introduced by
Mr. Elliott, the Commissioners engaged, for the purpose, the services
of Mr. Jay, now of the firm of Messrs. Quilter, Ball and Jay, most
experienced accountants in the city of London. We requested him, in
the first instance, to examine the books kept at Weedon during Mr.
Elliott's superintendence, and also those at present in use, with a
view of ascertaining the respective merits and defects of the two
systems. Messrs. Quilter and Co. have examined the books in use at
Weedon since December, 1855 (upwards of 300 in number) and have
made their report of the system upon which they have been kept.

It appears from that report that the bookkeeping at Weedon
between December 1855 and August 1856 was of a very rough and
imperfect description, and that there was no regular classification of
the facts indicated by the original vouchers. The books then kept
referred:

1. To the registration of papers and correspondence (these are
not strictly speaking books of account);

2. To dealings with contractors and tradesmen in connection
with the receipt of stores;

3. To receipts of stores from Woolwich, the Tower and other
government stations;

4. To orders for issues to regiments and issues made accordingly.

The accountants add, "Besides the books described above, a
rough document was framed having the character of a store ledger,
that is, containing accounts opened for some of the different
descriptions of stores, with entries purporting to show the respective
receipts and issues under distinctive heads, but it was never duly
entered up and as a book of results is perfectly useless. In addition to

the foregoing, certain other books were kept by the inspectors and viewers in their respective store rooms; there was also a foreman's book of issues, and one for special issues of cloth, both of which were likewise kept in the store rooms.

We see no reason to doubt that the entries made in these books were intended truthfully to record, and did in the main accurately record, the amount of stores received and inspected, packed and issued in each branch of the depot. But our accountants are of opinion that "although these books existed and entries more or less continuous were made in them, not one was kept efficiently and completely, and there was, properly speaking, no systematic book keeping; the consequence being a state of arrears and confusion which in a greater or less degree continued to characterize the accounts of the department down to the time when Captain Gordon took charge of it in May 1858 and notwithstanding the improvements which Mr. Elliott himself introduced in August 1856."

From and after the 22nd August 1856, a more regular system of accounts was adopted, which continued in operation during the remainder of Mr. Elliott's superintendence. The business of the depot was divided into four branches:

1. The registry branch, including all correspondence, letters both inward and outward, and the registration of papers and documents generally.

2. The contract branch, comprising the whole course of dealing with contractors and tradesmen in respect of supplies furnished by them, from the receipt of the goods at the depot to the granting of the certificates on which payment was made.

3. Receipt and issue branch. The business under this head consisted of taking account of articles brought into store, other than those received from contractors and tradesmen, and of all issues out of store.

4. Store ledger branch.

The business of this branch was to collect in one record, viz., "the store ledger" on the one hand, all receipts of clothing and stores from every source, and on the other, all issues of clothing and stores, with the view of exhibiting in debtor and creditor form under the head of each article, as "boots," "caps," "tunics," "trousers" & etc., the periodical receipts and issues and the balance or stock remaining on hand from time to time.

After a searching and laborious investigation into the accounts, during which the whole of the issues for the entire period from December 1855 to May 1858 have been traced to their various destinations, as indicated in the accounts, the accountants reported "that all stores delivered to Weedon, or coming within the scope of its official responsibility, have been substantially accounted for." And that "with respect to the personal accounts with contractors, they were enabled, after careful examination, to report that, excepting in some few and trifling instances, not calling for special observation, they found them to be essentially correct, and that no other moneys have been paid to the contractors than such as they became entitled to receive in consideration of stores delivered." And, finally, "that nothing came before them in the course of their investigation to warrant the suggestion of fraudulent practices by the late principal military storekeeper in dealing with the stores confided to his administration."

In the opinion thus expressed, the Commissioners entirely concurred; and they continued: We have great satisfaction in expressing to your Majesty our belief, that whatever suspicions may have been naturally excited, there has been, as regards the stores at Weedon, no dishonest dealing whatever. Everything which has been paid for has been received, and no defalcation has taken place. To have arrived at this conclusion makes us regret less than we should otherwise regret the time, labor and money spent in this inquiry. Nothing, indeed, can be a greater proof of the confusion and arrears into which the store accounts had fallen than the fact that it has taken Messrs. Quilter, Ball and Jay eight months at least to complete the ledgers and to arrive at the judgment on them above expressed.

As regards the cash accounts of Mr. Elliott, they were entirely distinct from and unconnected with the store accounts. All payments for supplies delivered at Weedon were made by drafts upon the paymaster-general, no storekeeper being allowed to have anything to do with the money transactions relating to the supply of stores. But it was Mr. Elliott's duty as storekeeper to pay the weekly wages of the foremen and laborers employed at the depot. These payments were always duly made by him and entered in check and pay lists, as laid down in the ordnance regulations. He was also entrusted with the duty of paying the charges of carriers for goods delivered at Weedon. For these purposes he made monthly, as is the custom with all storekeepers, a demand for the sum necessary for the expenses of the following month. Each demand showed the amount remaining to his debit after payment of the expenses of the preceding month. These demands during 1857 were for about £1,000 a month.

His cash accounts, with the necessary vouchers to support them, were sent in quarterly, in accordance with the regulations, fourteen days after the expiration of the quarter. His quarterly account ending the 31st December 1857 was rendered accordingly, showing a small balance against him.

His monthly demands for January, February, March and April 1858, including a demand of more than £1,600 for carriage of stores were granted. But as he neglected to send in his quarterly account up to the 31st March 1858 within the fourteen days prescribed, he was peremptorily called on to do so, and in default of his furnishing it, the accountant general refused the imprest for £2,000 demanded by him for the month of May. Had the quarterly account ending March 1858 been sent in, it must have been at once discovered that carriers' bills for £1,639 in respect of which he had obtained imprests had not been paid by him.

Mr. Elliott had also previously received instructions to make up his cash account to the date of his leaving Weedon for the post of storekeeper at Dublin, to hand over the balance in his hands to his successor, Captain Gordon, and to produce his receipt for the balance.

On the 17th of May, Mr. Elliott borrowed the sum of £500 from a contractor, out of which he paid £250 for the weekly wages of the establishment, which he appears never to have allowed to fall into arrears. Shortly after his disappearance on the 22nd of May, it was discovered that bills for the carriage of stores, amounting to £1,639.13s.4d. were unpaid by him, and that (including this sum) the balance due to the public from him was £2,048.10s.6d.

In accordance with the provisions of the Statute 52 Geo. III. c. 66, and the Ordnance Regulation No. 99, Mr. Elliott, on entering his office at Weedon, gave security in a bond for £2,000, entered into by a guarantee association on his behalf for the due performance of his duties. This £2,000 has since been paid by the guarantee association, so that the actual loss in cash occasioned by Mr. Elliott's deficiencies is reduced to £48.10s.6d. His cash accounts appear to have been kept with entire accuracy and by the proper system of double entry.

Upon a review of the whole evidence, the commissioners expressed their opinion that the general mode in which the business of the Weedon establishment was conducted was far from satisfactory.

We have specified, perhaps in tedious detail, the main defects and irregularities which existed there. The principal blame, which we can attach to Mr. Elliott, apart from the admitted deficiency in his cash balance, is in respect of his frequent absences from Weedon. No doubt it was often necessary for him to attend at the War Office and Mark Lane. But there is too much reason to believe that much of the time during which he was absent was devoted to his private pleasures in neglect of his public duty. When at his post, he seems to have worked diligently.

We acquit him of any deliberate intention to do wrong either in his so-called deviations from the ordnance system of accounts or in not keeping distinct the inspection and custody of the stores. We think that he was not sufficiently peremptory in insisting upon having the further help which he required.

In our opinion, the main defects in the Weedon establishment are chargeable to the War Department.

1. It was a mistake to fix the clothing depot so far from London, beyond the opportunity of immediate and frequent personal surveillance by the director and assistant director of clothing and separated from the important branch establishment in Mark Lane. This mistake has now been remedied by the abandonment of Weedon as a depot for clothing and the removal of the establishment to Pimlico.

2. It was a mistake to appoint the first head of the establishment without specific instructions as to its organization.

3. It was a great mistake to supply him at the outset of the undertaking with a small and inexperienced staff. The inspection staff in particular ought to have been organized from the commencement upon the plan adopted at the Tower and better salaries should have been given to the various inspectors.

4. It was a continuing mistake not to increase the staff in proportion to the increase of the work in accordance with Mr. Elliott's frequently repeated requests.

5. It was a mistake, notwithstanding the distance of Weedon from London, that more frequent visits were not paid there by the

directors of clothing, and a more rigid supervision exercised over the books and the stores.

6. It was a mistake not to have had a more considerable store in hand before commencing the issues, so as to have been prepared for the emergencies which arose.

7. It was, we think, a most serious mistake to have separate contracts for the cloth required, and for the making that cloth into garments; a course involving double contracts, double correspondence, double inspection, double carriage and double keeping of accounts. This defect has been cured, the contracts being now in the first instance entered into for the clothing completely made up.

These defects, together with the absence of proper patterns, the discrepancies between patterns and specifications, the haste with which tenders were called for, the delay in informing the storekeeper of the contracts entered into, the delivery of kits at Weedon, all these have satisfied us that the new system of clothing the army, however advantageous in many respects, was inaugurated without due consideration and certainly without adequate provision for so extensive a change. Neither the arrangements made nor the agents employed in making them were sufficient for the vast amount of labour which the new system required.

The laudable object of the War Department doubtless was to add the recommendation of economy to those to which their new system might be otherwise entitled. With this view, the establishment was stinted and the inspectors were miserably underpaid. Officers of intelligence, ability and practical experience received salaries of only £100 a year, a sum quite insufficient for the decent support of themselves and of their families. Yet to them was entrusted the power to decide upon the acceptance or rejection of goods of very large value, which contractors naturally wished not to have returned upon their hands. The result was inevitable.

These officers were thereby subjected not only to the suspicion that their services were considered of little value by their superiors, but to the graver imputation that persons so inadequately remunerated might be ready to show favor from corrupt motives to any contractor who could purchase their good will. We think there is no ground for such imputations.

Partial Text of the Report of the Investigation of
Weedon Bec Depot

We acquit not only Mr. Elliott, but all the officers of the War Department, as well as the inspectors and subordinate officers at Weedon, of having shown partiality or favor towards any contractor, and we have much pleasure in adding that a suggestion made before the Committee of the House of Commons on Contracts implying that favoritism had been shown to a particular contractor in respect of cloth delivered at Weedon was frankly withdrawn before us by the witness who had made it. We think there was ground for some suspicion in the first instance, but the explanation given of the facts cleared up the points, removed the impression on the mind of the witness himself and entirely satisfied us that no undue preference had been shown.

APPENDIX B
NORTHAMPTON SHOE FACTORY 1869

We see shoemakers and shoemakeresses at work in dingy ground floor rooms and at open upper windows; we noticed "Riveters' Entrance", & etc., painted on the finger-rubbed doors of the many windowed factories which might be taken for little cotton mills (metal rivets were used in cheaper kinds of footwear to attach soles to uppers and insoles).

Let us go to Messrs. Turner Bros (Hyde & Co)... in Campbell Square. The first impression produced us one of the queer contrasts that there are in the cordwainer's trade. The cobbler, cramped in his cupboard-like stall, belongs to it and so does the firm, which employs 400 hands on and four times as many off the premises. In one long room, five rows of clickers, with pale faces and dirty aprons with a penthouse or brief upper skirt of leather at the waist, are cutting on wooden slabs and blocks like butchers' all kinds of women's materials; in another, tougher men's materials are being manipulated. When cut, the uppers are rolled up, placed in ticketed baskets and sent up to the operatives in other parts of the premises or away to outside hands. A boot or shoe often goes out in this way twice before it is finished and stacked in the drying-room heated by steam pipes.

Down below, there is a puff of steam; wheels whir, bands run round and round, machinery clanks. Soles and heels and "split-lifts" (split-lift—narrow strip of leather wedge-shaped in section, curved so that it forms a marginal heel layer) are being punched out by iron frames that come down upon the leather with a thud, and when punched, slide down chutes into the bin-like receptacles beneath. These lads are pricking holes for the riveters by the aid of a machine; that old man is passing leather, to harden it, between steam-turned rollers.

It is curious to note the difference between hand work and machine work. Close by a sole-cutting machine, a young man or two are cutting up odds and ends of leather into soles by hand. Although they have the aid of the machinery to press the leather into shape, it is almost ludicrous to remark how few they make in comparison to the machine.

Soles and heels are garnered in great pigeon holes. Shaped leather of all sorts is arranged on shelves in ticketed baskets. Cistern-like tin-lined cases, inner lined with brown paper, are gaping for their loads.

Here is a pile of boots done up in pairs in white and green tissue paper; there is a pyramid of bright pink boxes, each holding a dozen pairs. Here the patent-leather tops of boots for South American gallopers over the Pampas are being eyeleted. Specially gay and graceful are the women's boots intended for Spanish-American countries; sky blue, with a golden star on the instep; mauve, golden-bronze, like a butterfly's wing, green, with a sheen like a drake's neck; pink, yellow and black, with coquettish little ankle tassels. Close by are shoes for New Zealand servant girls that looked as if their wearers would never need a second pair; and not far off, as substantial seeming sea boots for Newfoundland cod fishers. In an adjoining room there is an "infinite variety" a dazzling variety of many-colored babies' shoes, varying in price from 5d. up to 30s.

"And what is the value of Northampton's export of shoes?" "A million sterling per annum would be a low estimate" is the answer.

"And what are the average wages of the hands?" "Oh, it is almost impossible to strike an average. Some of mine, a very few, make £3 per week; more make £2; but I dare say a good many do not make more than 12s. It depends entirely on the man himself."

A large proportion of the Northampton shoemakers struck me, during my recent visit to them, as being decided members of the alcoholic persuasion. I met them mooning about, unshorn, unkempt – a condition in which too many of them remain in the day on which they need not work with filmy eyes which showed that they had gone on the "fuddle." I met them staggering. I saw them sparring one with his apron down, and the other with his apron hastily rolled up around his waist and then suddenly knocking off, knocking each other and amicably nodding their heads together, as if they had quite forgotten that they had been trying to blacken each other's eyes two seconds before. It is, I am informed, "the thing" with the Northampton shoemaker to take what he calls a "Sunday-Monday:" Id est, he works on Sunday, that he may have the more to lush on Monday.[1]

APPENDIX C
BRITISH FIRMS THAT CONDUCTED BUSINESS WITH THE CONFEDERATE STATES OF AMERICA

Aberdeen, Scotland
J&J Crombie & Co.

Belfast, Northern Ireland
B.&E. McHugh

Birkenhead
Birkenhead Iron Works
Laird Brothers

Birmingham Small Arms Trade (BSAT)
Cooper & Goodman
Pryse & Redman
Joseph Wilson
Joseph Bourne
EC Hackett
Thomas Turner
WL Sargant & Son
Bentley & Playfair
Joseph Smith
King & Phillips
R&W Aston
Swinburn & Son
Charles Maybury
W. Scott & Sons
Isaac Hollis & Sons
William Tranter
CW James

Robert Hughes
Ward & Son
E. Bond

Birmingham
Barner & Sons
Chatwin & Sons
WH Dowler
Firmin & Son
J.R. Grant & Son
S. Buckley & Co.
Smith & Wright Ltd.
Smith, Kemp & Wright
WM Middlemore

Clydebank, Scotland
John Brown & Co.

Dumbarton, Scotland
Denny Brothers

Glasgow, Scotland
Chamberlain & Co.
Clyde Bank Foundry
Patrick Henderson & Co.
James & George Thompson

Kelvinhaugh, Scotland
Alexander Stephen & Sons

Leeds
Greenwood & Batley Ltd.

Liverpool
Ashbridge & Co.
Blakely Ordnance Co.
Gordon Coleman & Co.
Collie, Westhead & Co.
Curry, Killoch & Co.

Cyclops Steel & Iron Works
Fawcett, Preston & Co.
Fletcher, Hall & Stone
George Forrester & Co.
Edward Lawrence & Co.
Leech, Harrison & Forwood
Low Moor Iron Works
William C. Miller & Son
Old Tug Co.
J. Stewart Oxley & Co.

London
Alexander Ross & Co.
Albion Trading Co.
Alexander Collie & Co.
Blakely Ordnance Co.
Curtis & Harvey
Christy's Hatmakers
De La Rue & Co.
Dudgeon Brothers
Emile Erlanger & Co.
Galway Co.
Graysbrook
John K. Gilliat & Co.
Overed Guerney& Co.
Hebbert & Co.
T.&C. Hood
S. Isaac Campbell & Co.
Sinclair, Hamilton & Co.
John Lane, Hankey & Co.
W.S. Lindsay & Co.
Zachariah C. Pearson & Co.
Railway Carriage Makers Railway Works
RT Tait & Co.
Robinson & Cottum
J. Henry Schroder & Co.
S. Straker & Sons
William & Co.
William Essex & Sons

Charles William & Co.
F. Wentworth & Co.

London Commercial Gunmakers
Potts and Hunt
Parker Field & Sons
J.E. Barnett & Sons
EP Bond
Freed & Co.
James Yeomans
London Armoury Co.
Keen & Son
Wilkinson
Holland & Holland

Manchester
Hammond, Turner & Bates
Lomnitz & Co.
Manchester Ordnance & Rifle Co.
Joseph Whitworth & Co.

Newcastle-Upon-Tyne
W.G. Armstrong & Co.
Elswick Ordnance Works

Northampton
Turner Bros, Hyde & Co.

Paisley, Scotland
J&P Coates

Sheffield
William Butcher, Jr. & Co.

Yorkshire (County)
Joshua Ellis & Sons[1]

Appendix D
Huse's British Imports

In early 1863, Gorgas compiled a statement of the supplies received from Huse. Of the £1,068,722 ($5,343,610) worth of supplies, nearly £600,000 ($3,000,000) came from SIC & Co.[1]

Ordnance Department

131,129 stands of arms as follows:
70,980 long Enfield Rifles
9,715 short Enfield rifles
354 carbine Enfield rifles
27,000 Austrian rifles
21,040 British muskets
20 small bore Enfield rifles
2,020 Brunswick rifles
At a cost (including cases, molds, kegs, screw drivers & etc.) of: £417,263,9s.11d ($2,086,315)

129 cannon as follows:
51 6-pounder bronze guns, smooth
18 howitzer bronze guns, smooth
6 12-pounder iron guns, rifled
2 howitzers, iron and carriages and caissons for same
6 rifled Blakely cannon, 6 3.10 inch carriages and 18,000 shells for same
2,000 fuses
3 rifled cannon
8 inch Blakely and 680 shells for same
12 rifled steel guns, 12 pounders and shot, shell & etc. for same
2 bronze guns, rifled
200 shells and fuses
756 shrapnel shell, round
9,820 wooden fuses

4 steel cannon, rifled, 9 pounders and 1,008 shells and fuses for same
220 sets harness
Spare parts artillery harness & etc.
At a cost of: £96,746,1s.8d ($483,730)

1,226 cavalry equipment
16,178 cavalry sabers
5,392 saber belts
5,392 saber knots
1,360 cavalry humnals (sic)
1,386 cavalry surcingles and pads
At a cost of: £20,321,12s.3d ($101,605)

54 sets web harness
456 leather butts
198 leather packages
At a cost of: £9,717,11s ($9,712)

34,731 sets of accoutrements
34,655 knapsacks complete (including mess tins and covers)
81,406 bayonet scabbards
4,000 canteen straps
40,240 gun slings
650 sergeants' accoutrements
At a cost of: £54,873,16s.3d ($274,365)

357,000 pounds of cannon powder
94,600 pounds of musket powder
32,000 pounds of rifle powder
900 pounds of bursting powder
4,137,000 cartridges for small arms
2,800 pounds of chlorate potassa
1,024 hundred weight of saltpeter
89,900 friction tubes
10,100,000 percussion caps,
At a cost of: £47,010,10s.3d ($235,050)

Appendix D

Quartermaster's Department:

74,006 pairs of boots
At a cost of: £28,422,16s.4d ($142,110)

62,025 blankets
At a cost of: £23,903,2s.11d ($119,525)

8,250 pairs of trousers
At a cost of: £5,144,11s.3d ($25,720)

170,724 pairs of socks
At a cost of £9,292,18s.7d ($46,460)

78,520 yards of cloth
At a cost of £24,660,15s.5d ($123,300)

703 shirts
At a cost of £738,9s.8d ($3,690)

8,375 greatcoats
At a cost of £13,294,17s.8d ($66,470)

17,894 yards flannel
At a cost of £1,632.5d ($8,160)

97 packages trimmings
At a cost of £3,435,11s.6d ($17,175)

Total cost of clothing: £110,525,3s.9d ($552,625)

Medical supplies
At a cost of: £13,482,10s.7d ($67,410)

46 sets of armorers tools
36 sets of saddlers tools
10 sets of farriers tools
3,336 pieces of serge for cartridge bags
2,000 cartridge bags

1,013 hundredweight lead
100 hundredweight sheet copper
16 flags
87 tarpaulins
10 hundredweight shellac
1,192 boxes of tin plate
75 packages of steel
64 hundredweight of steel
Total cost (including the medical supplies) £33,049.6d ($165,245)

Freight, railway carriage & etc.
At a cost of: £49,683,19s.5d ($248,415)

Supplies that have been shipped to date:

Small arms:	£417,262,9s.11d
Artillery and harness:	£96,746,1s.8d
Accoutrements & etc.	£54,973,10s.3d
Ammunition:	£47,010,10s.3d
Leather:	£9,717 11s
Clothing:	£110,525,3s.9d
Medical supplies:	£13,432,10s.7d
Ordnance stores:	£19,616,15s.5d
Freight, railway carriage & etc.	£19,732,7s.9d
Insurance & etc.	£29,951,11s.9d
At a cost of:	£818,869,18s.3d
	($4,094,245)

Supplies now in London ready for shipment

23,000 rifles to be delivered at Nassau	£87,950
20,000 scabbards	£1,500
46 casks of saddlers material	£631,5s
11 cases of nitric acid	£38,4s.8d
2,012,000 cartridges	£ 5,533,5s
3,000,000 percussion caps	£681
10,000 pouch tins for accoutrements	£250
286 ingots of tin	£628,2s.6d
931 of pigs lead	£1,252,17s.9d

Appendix D

3 cases thread, etc.	£240,17s.3d
1 bale of serge	£6019,6s
13,750 pairs of trousers, QM Dept.	£8,565,2s.1d
14,250 greatcoats, QM Dept.	£23,835,13s.17d
1,804 pairs of boots, QM Dept.	£887,11s.6d
4 chests of tea, Medical Dept.	£48,7s.6d
For a cost of:	£249,853,10s
	($1,249,265)
Total amount shipped:	£818,869,18s.3d
	($4,094,345)
Supplies to be shipped	
by December 15, 1863	£249,853.10
	($1,249,265)
Grand Total:	£1,068,722
	($6,246,325)

Appendix E
Subscriber List for the Erlanger Loan

Sir Henry de Houghton, Baronet	£180,000 ($900,000)
S. Isaac Campbell & Co.	£150,000 ($750,000)
Thomas Stirling Begbie	£140,000 ($700,000)
The Marquis of Bath	£50,000 ($250,000)
James Spence	£50,000 ($250,000)
Mr. Beresford Hope	£40,000 ($200,000)
George Edward Seymour	£40,000 ($200,000)
Msser Fernie	£30,000 ($150,000)
D. Forbes Campbell	£30,000 ($150,000)
Alexander Collie & partners	£20,000 ($100,000)
Patten Fleetwood/Wilson Schuster	£20,000 ($100,000)
W.S. Lindsay	£20,000 ($100,000)
Sir Coutts Lindsay, Baronet	£20,000 ($100,000)
John Laird, MP	£20,000 ($100,000)
M.B. Sampson	£15,000 ($75,000)
John Thaddeus Delane	£10,000 ($50,000)
Lady Georgina Fane	£15,000 ($45,000)
J.S. Gilliat	£10,000 ($50,000)
George Peacock, MP	£5,000 ($25,000)
Lord Wharncliffe	£5,000 ($25,000)
W.H. Gregory, MP	£4,000 ($20,000)
W.J. Rideout	£4,000 ($20,000)
Right Hon. William Gladstone	£2,000 ($10,000)
Edward Akenroyd	£1,500 ($7,500)
Lord Campbell	£1,000 ($5,000)
Lord Donoughomore	£1,000 ($5,000)
Lord Richard Grosvenor	£1,000 ($5,000)
Hon. Evelyn Ashley	£500 ($2,500)

APPENDIX F

ALEXANDER COLLIE & COMPANY

On July 4 1864, McRae included in his letter to Seddon the following enclosure, dated London, June 18:

> "*Memorandum of agreement between Alexander Collie of London, on the one part, and Colin J. McRae, as representing the government of the Confederate States of America, on the other part.*
>
> "*1. Alexander Collie agrees to provide four large and powerful new steamers to carry out the following arrangement with the least possible delay.*
> "*2. Alexander Collie will at once cause to be purchased, under Colin J. McRae's direction, quartermaster's stores to the value of 150,000 pounds sterling, and ordnance or medical stores to the value of 50,000 pounds sterling – the one subject to the inspection of Maj J.B. Ferguson, the other to that of Maj C. Huse.*
> "*3. The delivery of such purchases to extend over a period of about six months, in proportionate quantities, and shipment to be made to the Confederate States of America with as little delay thereafter as practicable.*
> "*4. Inland carriage and packing expense to be charged in the invoice and 2-1/2 percent commission to be charged also.*
> "*5. Colin J. McRae, on behalf of his government, agrees that on arrival in the Confederacy of any goods purchased and shipped by Alexander Collie under this agreement, such goods will be immediately claimed and taken over by the government. Fifty percent advance will be added to the English invoice, and Alexander Collie, through his agent, will immediately receive in exchange cotton at the rate of six pence per pound.*

"6. Such cotton to class middling and to be delivered alongside the steamers as required, compressed, packed and in good merchantable condition.

"7. Full cargoes of cotton received in exchange for goods delivered under this agreement may be shipped by Alexander Collie, through his agent, free from any other charge or restriction whatsoever beyond the now existing tare of one-eighth of a cent per pound.

"8. No steamers to have priority in any way over those employed by Alexander Collie in this service, and more than the four above mentioned may be used if Alexander Collie can arrange to put them on.

"9. Colin J. McRae further agrees that to cover the expense of Alexander Collie's agencies abroad, he, Alexander Collie, is to have the privilege of providing and bringing out other cotton than that received under this agreement to the extent of one-tenth part of the cargo space of the respective steamers, and such cotton (or tobacco) may be shipped on the same terms as indicated per government cotton, viz, free from all other charges or restrictions whatsoever excepting the before named export duty now existing.

"10. This agreement is to be construed by both parties in a spirit of confidence and liberality. The one will purchase and send forward the supplies indicated with the least possible delay; the other will deliver cotton as required in the same way, and neither party will withhold necessary supplies on account of any temporary shortcoming on the part of the other.

"11. Alexander Collie's agents, with the necessary staff for attending to this business, are to be allowed the privilege of residing in the Confederacy free from liability to conscription, and every reasonable facility is to be allowed them for effectually carrying out the terms of this agreement
"Alexander Collie,
"C J McRae, Agent for Confederate States of America."[1]

★ ★ ★

Alexander Collie was born in Aberdeen, Scotland, in 1823, the son of a merchant. He started a cotton import/export business in Manchester in 1850. By 1863, he had established a warehouse in London. His brother, William, ran the business in Manchester, while Alex concentrated on the London side of the enterprise. Another brother, George, ran a separate business in Liverpool, known as George Collie & Co. and should not be confused with Alexander's and William's business.[2]

On his arrival in England, William Crenshaw pitched the idea of a joint venture for the creation of a private steamship line, of which half the cargo would be for the Confederate War Department. The venture would create a monopoly on all quartermaster and commissary stores run through the blockade and a commission on the shipped goods, whether the ships made it to their port of destination or not. Collie advanced the money for the first steamship, the *Havana*.

The initial appeal to investors was based on the assumption, which proved false, that the Union blockade would not disturb commercial vessels filled with quartermaster and commissary goods, which could be classified as non-contraband under international law. In contrast, Ordnance Department supplies were always classified as contraband and thus subject to seizure as a prize of war.[3]

Crenshaw and Collie had joint interests in some steamships and not in others. Collie had a side deal going on with the state of North Carolina for proprietary supplies backed by their own cotton, which did not involve supplies for Richmond. Since states rights was a key provision of the reason behind secession from the Union, state-owned vessels were exempt from the general requirement that half the cargo space on merchant vessels be reserved for the Confederate government.

A series of tragedies began to befall Crenshaw. First, in mid-May 1863, his Richmond factory burned to the ground, leaving only his steamship enterprise as a source of income. Then, over a short period of time, the Crenshaw steamships began to suffer a disproportionate number of captures compared to those being operated separately by Alexander Collie & Co. Crenshaw lost seven steamships in the summer of 1864, all on either their first or second voyage, while

Alexander Collie & Co. enjoyed a much more successful record, including those shipments for North Carolina.

At the time, the port of Wilmington was the second most important city in the Confederacy after Richmond. Wilmington was running almost three times the volume of cargo of Charleston. Alexander Collie & Co. maintained a staff of agents in Wilmington to handle its business interests in the city.

Following the collapse of the Confederacy, Collie tried to keep his firm afloat. In 1875, the firm finally collapsed in an enormous financial scandal. Collie and William had obtained large sums of money on false pretenses. The London and Westminster Bank believed that certain bills used were genuine trade bills, accepted by Collie's customers in respect of genuine transactions. In fact, the bills were accommodation bills which held no value at all. Both brothers were immediately arrested and spent two nights in Newgate Jail before being released on bail. On August 9, Collie failed to surrender to the terms of his bail conditions and fled to Spain, which at the time did not have an extradition treaty with England.[4]

The following account summarized the scandal:

> *"In the middle of June, 1875, disaster again occurred this time in the Manchester and India trade by the failure of Alexander Collie & Co., of Manchester and Leadenhall Street, London, and, in consequence, this year became known then and since, as the 'Collie' year. The liabilities of this firm were estimated at three million; but firm after firm, as a result of the failure of this house, suspended payment, one house having nominal liabilities of two and a-half million.*
>
> *"The firm of Collie & Co. consisted of two brothers, Alexander and William Collie. The trustee in bankruptcy was the late Mr. John Young, of Turquand Young, and the realization of the Estate proved to be a lingering and disastrous one for the Creditors, the actual amount of dividend paid by this particular estate being disbursed in six installments, in 1876, 1878, 1880, 1883, 1885 and the final closing of account took place in June 1889. At the meetings of Creditors, and by statements from*

responsible quarters, it transpired that the debtors had been living in all the prodigality of luxury that, between them, they had been in the habit of drawing 20,000 a year for their personal expenditure, Mr. Alexander Collie having drawn 123,000 for himself.

"The lease of Alexander Collie's house in Kensington Palace Gardens sold for 38,500, and the 'costly contents' were the subject of five days' sale. It is notorious in connection with that year that, as a consequence of losses sustained; the leading Joint Stock Banks had to reduce their half-yearly dividends, in addition to otherwise providing for the losses sustained.

"A feature of this firm's transactions was, that they did not accept Bills, and there was not a single Acceptance of theirs existent at the date of their failure. They drew upon Houses, many of whom were entirely of their own creation, and who were financed by them. On the 21st July, both Partners, whose Capital existed only in name, were charged at the Guildhall Police Court for having obtained money by false pretences, in drawing Bills with marks and numbers upon them, indicating that they referred to Cotton and Ledger accounts; but in fact, they were only Accommodation Bills.

"It was proved that there were no such goods sold, no such accounts in the ledger, and no goods accounts between the acceptors and the guarantors. As a matter of fact, goods to the extent of 100,000 roughly were all that were represented in millions of Bills in the hands of Banks and other Firms.

"On the 8th August, it was announced that Alexander Collie had absconded. Warrants were granted for his apprehension, but he was not arrested, and it was afterwards ascertained that he was in Spain, between which country and this, at that time, no extradition treaty existed. The charge was not proceeded with as against William Collie. Alexander Collie died in New York on 23rd November, 1895. The effect of this failure was severely felt in the Manchester and India goods markets."[5]

Appendix G
U.S. Supreme Court Decision
on the *Springbok*

The *Springbok*, 72 U.S. 1 (1866)

Syllabus

1. Though invocation, in prize cases, is not regularly made on original hearing, but only after a cause has been fully heard on the ship's documents and the preparatory proofs, and where suspicious circumstances appear from these, yet where the court below, in the exercise of its discretion, has allowed it on first hearing, the decree will not necessarily be reversed, decrees of condemnation having passed in both the cases invoked, one pro confesso and the other by a decree of the highest appellate court.

2. Where the papers of a ship sailing under a charter party are all genuine and regular and show a voyage between ports neutral within the meaning of international law, where there has been no concealment nor spoliation of them, where the stipulations of the charter party in favor of the owners are apparently in good faith, where the owners are neutrals, have no interest in the cargo, and have not previously in any way violated neutral obligations, and there is no sufficient proof that they have any knowledge of the unlawful destination of the cargo, in such a case, its aspect being otherwise fair, the vessel will not be condemned because the neutral port to which it is sailing has been constantly and notoriously used as a port of call and transshipment by persons engaged in systematic violation of the blockade and in the conveyance of contraband of war, and was meant by the owners of the cargo carried on this ship to be so used in regard to it.

3. The facts that the master declared himself ignorant as to what a part of his cargo, of which invoices were not on board (having been sent by mail to the port of destination) consisted, such part having been contraband, and also declared himself ignorant of the cause of

capture, when his mate, boatswain and steward all testified that they understood it to be the vessel, having contraband on board, held not sufficient, of themselves, to infer guilt to the owners of the vessel, in no way compromised with the cargo. But the misrepresentation of the master as to his knowledge of the ground of capture held to deprive the owners of costs on restoration.

4. A cargo was here condemned for intent to run a blockade where the vessel was sailing to a port such as that above described, the bills of lading disclosing the contents of 619 packages of 2007 which made the cargo, the contents of the remaining 1388 being not disclosed; where both they and the manifest made the cargo deliverable to order, the master being directed by his letter of instructions to report himself on arrival at the neutral port to H., who "would give him orders as to the delivery of his cargo;" where a certain fraction of the cargo whose contents were undisclosed was specially fitted for the enemy's military use and a larger part capable of being adapted to it; where other vessels owned by the owners of the cargo, and by the charterer, and sailing ostensibly for neutral ports were, on invocation, shown to have been engaged in blockade running, many packages on one of the vessels, and numbered in a broken series of numbers, finding many of the complemental numbers on the vessel now under adjudication; where no application was made to take further proof in explanation of these facts, and the claim of the cargo, libeled at New York, was not personally sworn to by either of the persons owning it, resident in England, but was sworn to by an agent at New York, on "information and belief."

Appeal from a decree of the District Court of the United States for the Southern District of New York respecting the British bark *Springbok* and her cargo, which had been captured at sea by the United States gunboat *Sonoma* during the late rebellion and libeled in the said court for prize.

The vessel was owned by May & Co., British subjects, and was commanded by James May, son of one of the owners.

She had been chartered 12th November, 1862, by authority of May, the captain, to T.S. Begbie, of London, to take a full cargo of "Lawful merchandise, and therewith proceed to Nassau, or so near thereunto as she may safely get, and deliver same, on being paid freight as follows & etc., the freight to be paid one-half in advance on clearance from custom house, subject to insurance, and the remainder in cash on delivery.

"Bills of lading are to be signed by master at current rate of freight, if required, without prejudice to this charter party. It being agreed that master or owners have an absolute lien on cargo for all freight, dead freight, demurrage or other charges. The ship is to be consigned to the charter's agent at port of unloading, free of commission. Thirty running days are allowed the freighter for loading at port of loading and discharging at Nassau."

This document had an endorsement on it by Speyer & Haywood, persons hereinafter described.

The letter of instructions to the master was thus:

> "*London, December 8, 1862*
> "*Captain James May*
>
> "*Dear Sir:*
>
> "*Your vessel being now loaded, you will proceed at once to the port of Nassau, N.P., and on arrival report yourself to Mr. B.W. Hart there, who will give you orders as to the delivery of your cargo and any further information you may require.*
>
> "*We are, dear sir & etc.*
> *Speyer & Haywood,*
> *For the Charters*"

The letter to the agent of the consignee directed to B.W. Hart, Nassau and from these same persons, Speyer & Haywood, was thus:

> "*Under instructions from Messrs. Isaac Campbell & Co., of Jermyn Street, we enclose you bills of lading for goods shipped per Springbok, consigned to you.*"

The London custom house certificate was "from London to Nassau;" the certificate of clearance declared the "destination of voyage, Nassau, N.P.;" and the manifest was of a cargo from "London to Nassau."

The log book was headed, "Log book of the bark *Springbok* on a voyage from London to Nassau."

The shipping articles, November, 1862, were of a British crew, "on a voyage from London to Nassau, N.P.; thence, if required, to any other port of the West India Islands, American ports, British North America, east coast of South America and back to the final port discharge in the United Kingdom or continent of Europe, between the Elbe and Brest, and finally to a port in the United Kingdom; voyage probably under twelve months."

The cargo, valued at £66,000, was covered by three bills of lading (of which two were duplicated, the duplicates marked Captain's copies), as follows:

Bill of lading marked Number 2 showed "666 packages merchandise," shipped by Moses Brothers, to be delivered, etc., at the port of Nassau, N.P., unto order or to assigns, he or they paying freight, as per charter party. It was endorsed by Moses Brothers in blank.

"This bill of lading on its face showed 150 chests and 150 half-chests tea, 220 bags coffee, 4 cases ginger, 19 bags pimento, 10 bags cloves, and 60 bags pepper – in all, 613 packages. The remaining 53 were entered as cases, kegs and casks. These 53 packages were found, when the cargo was more closely examined, to contain medicines and saltpeter, matters at that time much needed in the Southern states, then under blockade.

"Bill of lading Number 3 showed one bale and one case shipped by Speyer & Haywood, to be delivered at Nassau unto order or to assigns & etc., paying freight as per charter party.

"Bill of lading Number 4 showed 1,339 packages shipped by Speyer & Haywood to Nassau, as above. These 1,339 packages were also described as cases, bales, boxes and a trunk. This was also endorsed in blank.

"The manifest gave no more specific description of the character of the cargo. It was signed Speyer & Haywood, brokers, and showed that the whole cargo was consigned to order."

An examination of the packages in bills Numbers 3 and 4 showed 540 pairs of gray army blankets, like those used in the army of the United States, and 24 pairs of white blankets; 360 gross of brass navy buttons, marked CSN; 10 gross of army buttons marked A.; 397 gross of army buttons marked I; and 148 gross of army buttons marked C.;

being in all 555 gross; all the buttons were stamped on the underside Isaacs Campbell & Co., 71 Jermyn St., London.

There were 8 cavalry sabers, having the British crown on their guards; 11 sword bayonets, 992 pairs of army boots, 97 pairs of russet brogans and 47 pairs of cavalry boots, etc.

The vessel set sail from London, December 8, 1862, and was captured February 3, 1863, making for the harbor of Nassau, in the British neutral island of New Providence and about 150 miles east of that place. The port, which lay not very far from a part of the southern coast of the United States, it was matter of common knowledge had been largely used as one for call and transshipment of cargoes intended for the ports of the insurrectionary states of the Union, then under blockade by the Federal government. The vessel, when captured, made no resistance, and all her papers were given up without attempt at concealment or spoliation.

Being brought into the port of New York and libeled there as prize, February 12, 1863, a claim was put in on the 9th of March following by Captain May for his father and others as owners of the vessel. On the 24th of the same month, a claim for the whole cargo was put in for Isaac Campbell & Co., and also for Begbie through one Kursheet, their agent and attorney; Kursheet stating in his affidavit in behalf of these owners that "it is impossible to communicate with them in time to allow them to make the claim and test the affidavit herein." His affidavit stated farther, "That as he is informed and believes, it was not intended that the barque should attempt to enter any port of the United States or that her cargo should be delivered at any such port, but that the only destination of such cargo was Nassau aforesaid, where the said cargo was to be actually disposed of and proceeds remitted to said claimants.

"That, as he is informed and believes, the cargo was not shipped in pursuance of any understanding either directly or indirectly, with any of the enemies of the United States or with any person or persons in behalf of or connected with the so called Confederate States of America, but was shipped with the full, fair and honest intent to sell and dispose of the same absolutely in the market of Nassau aforesaid.

"That his information is derived from letters and communications very lately received by this deponent from the aforesaid claimants, and from documents in deponent's possession, placed there by said claimants, and that such communications authorize this deponent to intervene and act as agent as well as proctor and advocate for the said claimants as to the above cargo."

The master, mate and steward, were examined as witnesses *in preparatorio:*

The master stated that the goods were to be delivered at Nassau for account and risk of Begbie & Co., London, the charterers; that he did not know that the laders or consignees had any interest in the goods; that he knew nothing of the qualities, quantities or particulars of the goods or to whom they would belong if restored and delivered at the destined port; that he was not aware that there were goods contraband of war on board; that, as he believed, invoices and duplicate bills of lading were sent to Nassau by mail steamer; that there were no false bills of lading, nor any passports or sea-briefs other than the usual register and ship's papers, which were entirely true and fair; that he did not know on what pretence she was captured; that there were no persons on board owing allegiance to the United States; that on the vessel's previous voyage, she went from London to Jamaica, carrying general merchandise and returned direct, carrying principally logwood.

The mate, who to a greater or less extent confirmed these statements, swore that the cargo was a general cargo – casks, bales, boxes and bags; that he had no knowledge, information or belief as to what was contained in them and had never heard. He knew of no goods contraband of war; no arms or munitions of war that he knew of, "The seizure," he stated, "was made on the supposition that the cargo was contraband of war."

The boatswain testified to the same purpose of the voyage; that the vessel had no colors but English aboard; that the cargo was general, in bales, cases and bags, that he did not know their contents and never had heard them stated, and that he "understood the seizure was made because the bills of lading did not show what was in some of the cases on board."

The steward, that he "understood the vessel was captured because we had goods contraband of war aboard; had heard no other reason given."

Upon the hearing in the district court, the counsel for the captors invoked into the case the proofs taken in two other cases, on the docket of that court for trial at the same time with the present one, the cases, namely, of United States v. Steamer *Gertrude* and United

States v. Schooner *Stephen Hart*. The *Hart* was captured on the 29th of January 1862 between the southern coast of Florida and the Island of Cuba.

The claimants of her whole cargo were the firm of Isaac Campbell & Co., the same persons who claimed, jointly with Begbie, the cargo of the *Springbok*. It also appeared that in the case of the *Stephen Hart*, the brokers who had charge of the lading of her cargo were Speyer & Haywood, the same parties who appeared as brokers of the cargo in the present case, and as shippers of a part of it, and as agents for Begbie and for SIC & Co.

It appeared in the case of the *Stephen Hart* that SIC & Co. were dealers in military goods, and that the entire cargo of that vessel, consisting of arms, munitions of war and military equipments, was laden on board of her in England under the direction of SIC & Co., in cooperation with the agents, at London, of the Confederate States with the design that the cargo should run the blockade into a port of the enemy either in the *Stephen Hart* or in a vessel into which the cargo should be transshipped at some place in Cuba, and that SIC & Co. entrusted to the agent of the Confederate States in Cuba, the determination of the question as to the mode in which the cargo should be transported into the enemy's port. The cargo of the *Stephen Hart* had been condemned by the Supreme Court as lawful prize at the last term. U.S.

The *Gertrude* was captured on the 16th of April 1863 in the Atlantic Ocean off the Bahama Islands while on a voyage ostensibly from Nassau to St. John's, N.B. The libel was filed against her on the 23rd of April 1863, and she was condemned with her cargo as lawful prize on the 21st of July 1863. No claim was put in to either the *Gertrude* or her cargo. The testimony showed that she belonged to Begbie; that her cargo consisted, among other things, of hops, dry goods, drugs, leather, cotton cards, paper, 3,960 pairs of gray army blankets, 335 pairs of white blankets, linen, woolen shirts, flannel, 750 pairs of army brogans, Congress gaiters and 24,900 pounds of powder; that she was captured after a chase of three hours, and when making for the harbor of Charleston, her master knowing of its blockade and having on board a Charleston pilot under an assumed name.

Appendix G

The marshal's report of the contents of the packages on board the *Springbok* and of the prize commissioners' report of the contents of the packages of the *Gertrude* disclose the following facts:

The report in the case of the *Springbok* specified eighteen bales of army blankets, butternut color; each marked A, in a diamond, and numbered 544 to 548, 550, 552, and 555 to 565. The report in the case of the *Gertrude* showed a large number of bales of army blankets, each marked A, in a diamond, and numbered with numbers, scattered from 243 to 534, and then commencing to renumber again at 600.

In the cargo of the *Springbok* was found a bale marked A, in a diamond, and numbered 779; while in the cargo of the *Gertrude* were found bales each marked A, in a diamond, and numbered 780, 782, 784, 786, 788, 789 to 799.

In the *Springbok* were found nine cases, each marked A, in a diamond, and numbered 976 to 984, and 4 bales, each marked A, in a diamond, and numbered 985 to 987 and 989, by the same marks, the 4 bales being stated to be "men's colored traveling shirts."

In the *Gertrude* were found 5 bales, each marked A in a diamond and numbered 998, 990 to 992 and 998, and described as "men's colored traveling shirts." In the *Hart* were 4 cases of men's white shirts, each marked A in a diamond and numbered 994 to 997.

So also, in the *Springbok* were found packages each marked A, in a diamond, SIC & Co., and numbered irregularly and with considerable hiatus, from 1221 up to 1440. But there was no 1285 among them, the hiatus being from 1266 to 1289, which last was the first of several having shirts. On the *Gertrude* were packages marked A, in a diamond, numbered from 1170 to 1214, also one numbered 1285, and found to contain shirts.

On board of the *Springbok* was found one bale of brown wrapping paper, marked A, in a diamond, T.S. & Co. and numbered 264. On board of the *Gertrude*, a large number of bales of wrapping paper and other paper marked A, in a diamond, T.S. & Co. and numbered with numbers scattered between 1 and 170.

In only one instance, apparently, so far as the testimony showed, was the same number found on a package in each cargo. On the other hand, many marks were found on the one vessel not found on the other.

No application was made in the court below for leave to furnish further proofs.

The court below condemned both vessel and cargo.

Appendix H
Boots and Shoes Supplied
By SIC & Co.

**A British Army Blucher shoe exported to the
Confederate States of America.**
(Courtesy of the New York State Museum)

In February 1863, Gorgas reported that Huse purchased 74,006 pairs of boots for £28,422,16s.4d ($142,110). The boots supplied to the Confederacy were mostly copies of the British Army Blucher boot, especially made for the Confederate market.

Late in the Napoleonic Wars, the British Army issued lace-up ankle boots to replace the older buckle shoes. The boots were named after Prussian general Gebhard Leberecht von Blücher. Blücher commissioned a boot with side pieces lapping over the front in an effort to provide his troops with improved footwear. This design was adopted by armies across Europe including Britain.

Blucher boots remained in use throughout the 19th century and were used in the Crimean War (1853-1856), First Zulu War (1879) and First Boer War (1880-1881).

This style of boot was patterned after the half boot, which had an open throat with laces which tied over the tongue itself. The boot had two metal eyelets and a herringbone twill pull loop at the back of the ankle. They were made from black waxed leather and featured hobnails with a steel heel rim.

They were described in a period manual:

> *"They were clumped soled, the clumps being at once toothed along the top edges and decorated by a course of deep-headed brass nails. The toe-point was, as we call it, capped and the meeting parts were closed with a bold flat-seam. The front lacing holes were strengthened by lines of stab work, one to hold down the inside over-plying of the binding leather – a strip of brown calf grain and the other merely decorative. The laces too, were leather, well pressed and rounded, while the holes they were to work in were substantially protected with metal rings."[1]*

British imported buckle shoes
(Courtesy of the Military & Historical Image Bank)

The other type of shoe imported to the Confederacy in large numbers was the now outdated buckle and strap type. This style of boot had been in service in the British Army for decades before the introduction of the new lace-up Blucher. In April 1863, a Federal quartermaster, examining a captured pair of these shoes, described them as: "made of well-tanned leather very well curried, but not blacked on the grain side, is as usual high in the ankle and confined by straps and buckle instead of string. These are the best I have seen for army use."[2]

Both styles of these boots were manufactured by Turner Bros, Hyde & Co., which, by 1864, employed over 300 people and was producing 100,000 pairs of shoes and boots a week.

Northampton was the principle producer of footwear for the Confederate Armies and produced hundreds of thousands of pairs of shoes and boots for the Confederate market during the course of the war.[3] Most of these commercial copies of the regulation boots and shoes were produced in black leather, but SIC & Co. also sent thousands of pairs of russet bluchers to the Confederacy.[4]

When the *Springbok* was captured, she carried "992 pairs of army boots"[5] and "97 pairs of russet brogans."[6] The *Stephen Hart* carried "2,000 pairs of brogan shoes & 592 pairs of russet shoes, blucher pattern; 762 pairs of black leather shoes, blucher pattern."[7]

In late 1863, General Kirby Smith of the Trans-Mississippi Department sent Minter to England to see to the contracts already made and to purchase further quartermaster and ordnance stores. Minter teamed up with Ferguson and together they bought the following items from Turner Bros, Hyde & Co.:

> "November 12, 1864: 41 cases sewed army shoes, 4,100 pairs. 31 cases sewed army shoes, 3,100 pairs.
> "November 29, 1864: 21 cases army shoes, 2,100 pairs. 9 cases army shoes, 900 pairs.
> "December 19, 1864: 14 cases army shoes, 1,400 pairs."[8]

How many thousands of pairs of shoes and other equipment were sold by Turner Bros, Hyde & Co. to other Confederate Departments is not known, but it is clear proof that Samuel and Saul Isaac, indirectly at least, still managed to sell goods to the Confederacy despite the ban placed on them.

Apart from the P1853 rifle musket, shoes were the only items that were imported throughout the war. Five hundred forty-five thousand shoes shipped from England from November 1, 1863 to December 8, 1864.[9]

Add to this, the 74,006 pairs received by Gorgas in February 1863, and the shoes bought by Minter, and the total amount of British boots and shoes imported from the outbreak of the war until December 1864 was over 630,000 pairs.

A dead Confederate soldier in the trenches of Petersburg. Under a magnifying glass, the metal eyelets on his British imported Blucher boots can clearly be seen.

(Courtesy of the Library of Congress)

APPENDIX I
INSPECTION CERTIFICATE

Isaac Curtis Inspection Certificate
(Courtesy of The McRae Papers)

Isaac Curtis was employed by SIC & Co. as a viewer. His job was to inspect all the rifles and muskets bought by the company through their various suppliers to make sure they were of the best possible quality. After inspecting and checking the weapons, he stamped the gun with his own personal mark, usually his initials, and passed them as fit for export to the Confederacy.

On August 22, 1861, Anderson wrote in his diary: "Bought from Isaacs 10,000 muskets old pattern for the Confederacy."[1]

The next day, he contracted with SIC & Co. for "11,000 English muskets of very good quality."[2] These guns showed up on Gorgas' February 3, 1863 letter as "21,040 British muskets."[3]

The weapons mentioned above were in fact outdated surplus Pattern 1851 muskets bought by Anderson from SIC & Co. on August

22 and 23, 1861, and inspected by Isaac Curtis from September 25 through October 1861. (See certificate above)

These weapons, though out-dated, were better than those being considered by the Confederate government at this early stage of the war. Such was the dire need for any kind of arms that out-dated or non-percussion weapons were considered. Secretary of War Walker wrote to both Huse and Anderson on August 30, 1861, to stress the point:

> *"To meet the large forces our enemy is endeavouring to hurl against us, we must have additional arms before supplies can be obtained from our own factories, just going into operation. If you cannot do better, you had better procure and forward without delay the flint muskets mentioned by Captain Huse, with flints and ammunition for the same."*[4]

Curtis and another viewer named Hughes would inspect the British regulation P1853 Enfield rifle muskets secured by SIC & Co. from the London gun-maker J.E. Barnett & Sons in 1862.

On May 26, 1862, SIC & Co. sent an invoice for the sum of £46.17s,2d. ($230.00) to J.E. Barnett & Sons for guns inspected by Curtis and Hughes from January 4 to May 24, 1862.[5]

The IC of the Isaac Curtis Stamp
(From the David Burt Collection)

Appendix I

The IC mark is found stamped on the stocks of three Barnett P53 Enfield rifles in the collection of the Atlanta History Center. The mark consists of a small circle containing the letters IC. Also found on various J.E. Barnett & Sons' weapons is another stamp consisting of a small circle containing the letters CH over the number one. This mark has long been believed to be the stamp of Caleb Huse, but Huse would not have had the time or the expertise to have done this kind of work. It is probably one of the stamps used by Hughes. On at least one surviving Barnett Enfield is another mark, a small letter H found stamped below the trigger guard. The CH and the H marks are the initials for Hughes and not Caleb Huse.

Appendix J
Major General William Dorsey Pender's Trousers

Pender's Trousers
(From the David Burt Collection)

On display in the Museum of the Confederacy is an intriguing pair of British import trousers intended for use by enlisted men. They are made of royal blue kersey with pockets and inner facings of light brown cotton drill with black japanned tin buttons and a narrow waistband with reinforcements behind the trouser buttons. Most importantly, they have a label inside which is described as:

> "The label is paper glued to the kersey inside the pants. It looks to have originally read 'S. Campbell & Co.' There are two lines that look to have originally read "Waist" and "Length," next to which sizes were hand written in ink."[1]

The trousers bear no WD broad arrow acceptance stamp. This verifies that the trousers were not made for the British Army but were made by a SIC & Co. sub-contractor.

Pender was wearing these trousers when he was mortally wounded on the second day of the Battle of Gettysburg. There is a cut in the left leg of the trousers where the shell fragment entered and the wound tended. The leg was later amputated. Pender died from his wound on July 18, 1863.

APPENDIX K
BUTTONS SUPPLIED BY SIC & CO.

SIC & Co. never produced buttons, instead relying on button manufacturers – mainly the firm of Smith, Kemp and Wright of Birmingham. The only uniform items SIC & Co. provided in any quantity were greatcoats and uniform trousers for enlisted men and, in smaller quantities, frockcoats for the officers.

SIC & Co. provided at least ten different kinds of buttons to the Confederacy, including three different buttons for the Confederate Navy. All the buttons produced for SIC & Co. are stamped Isaacs instead of Isaac as in the S. Isaac Campbell & Co logo, which appears on the famed stamp on their leather articles.

The Hughes Trouser Button
(From the David Burt Collection)

The button pictured above is from the trousers of Major John Hughes, a quartermaster in the 7th North Carolina Infantry. The SIC & Co. supplied button was used on the waistband and fly. These buttons were also used on undergarments. The button is made of stamped brass with a concave center and four holes for sewing into the garment. The button is approximately 1/2" in diameter and stamped with the S. Isaacs Campbell & Co/London/71 Jermyn St stamp.

Other buttons on the trousers were made the same, but marked S. Isaacs Campbell & Co only. Buttons from these trousers are the rarest of all buttons supplied by SIC & Co.

Infantry Buttons

Used SIC & Co. Infantry Button
(From the David Burt Collection)

Back of the Button
Showing the SIC & Co. Back Mark
(From the David Burt Collection)

The button in the above photograph is an excavated Old English manuscript lined I button, size 7/8". It is made of two pieces of brass and has a fixed shank at the rear. The inner surface of the letter itself is made up of fine horizontal lines.

This button was imported by the hundreds of thousands; indeed in just one shipment on the *Justitia*, which set sail on November 11, 1862, "2,000 gross large brass I buttons,"[1] which equals to approximately 288,000 infantry buttons were shipped.

In early 1863, the *Springbok* carried "555 Gross Army Buttons" (79,920).[2] These buttons also carried the S. Isaacs Campbell & Co/71 Jermyn St/London back mark as faintly seen in the picture.

This style of button was used on countless Confederate enlisted men's and officers' coats, including the uniform coat of Colonel Laurence Massilion Keitt of the 20th South Carolina Infantry, which is now displayed in the Museum of the Confederacy in Appomattox Courthouse, Virginia.

Cavalry Buttons

Cavalry buttons were exactly the same style as the infantry buttons, but carried the Old English letter C for cavalry. The buttons were two pieces, made of brass, with the Old English lined C and fine horizontal lines within the letter itself. The button is 7/8" in diameter or the same size as the infantry button and carried the S. Isaacs Campbell & Co/71 Jermyn St/London back mark.

The *Justitia* carried 300 gross "Large brass C buttons," which equates to approximately 43,200 buttons.[3] When the *Springbok* was captured, she had 148 gross (21,312) of C marked buttons.

The buttons were back marked with S. Isaacs Campbell & Co/71 Jermyn St/London.[4]

Artillery Buttons

The company also produced the Old English A buttons for the artillery. The letter had within it horizontal raised lines, was back marked as the other buttons and was 7/8" in diameter.

The *Springbok* carried 10 gross (1,440) of A buttons, but the *Justitia* successfully delivered 500 gross (72,000) "large brass A buttons,"[5] in December, 1862. All total, the *Justitia* transported over 403,200 large brass SIC & Co. marked buttons, enough to outfit approximately 50,000 men. (If the average jacket carried eight buttons as the Richmond Depot jackets did) This did not include buttons – as detailed below – for the other branches of service within the Confederate Armies.

Engineers Buttons

Manuscript E (Engineers) buttons were also made for SIC & Co. These were larger button than the ones produced for the infantry, cavalry and artillery, being 1" in diameter. This time the buttons were back marked S. Isaacs Campbell & Co/71 Jermyn St/London.

General Staff Buttons

General staff buttons consisted of two pieces, convex, bearing an eagle design with a large shield inscribed CSA and 11 six-pointed stars around the edge. The button was an inch in diameter and back marked S. Isaacs Campbell & Co/71 Jermyn St/London.

These buttons were used on Confederate officer uniforms, including the uniform coat of Captain Marshall Hairston of the 15th Mississippi Infantry, Major General Henry DeLamar Clayton of the Army of Tennessee, Brigadier James Morris Hawes of the Trans-Mississippi Department and the frock coat of Major John Hughes.[6]

Navy Buttons

The buttons supplied to the Navy were:

Crossed cannons with fouled anchor design, 7/8" in diameter and marked CSN. The button was stamped with S. Isaacs Campbell & Co/71 Jermyn St/London and back marked S. Isaacs Campbell & Co/St. James St/London.

Crossed cannons with fouled anchor design, 5/8" in diameter and marked CSN. The button was back marked S. Isaacs Campbell & Co/71 Jermyn St/London.

Crossed cannons with fouled anchor design, 5/8" in diameter and marked CSN. The button was back marked 71 Jermyn St.

Three hundred sixty gross (51,840) Navy buttons, marked CSN[7], back marked S. Isaacs Campbell & Co/71 Jermyn St/London were on the *Springbok*.

General Service Buttons

Although no general service buttons bearing the CSA stamp have been located with the S. Isaacs Campbell & Co. back mark, at least one shipment with this CSA design was destined for the Confederacy. They were on board the *Stephen Hart* when she was captured off the coast of Florida on January 29, 1862.

In the District Court's decision on the *Stephen Hart* read:

"One important circumstance, to show that the cargo of the Stephen Hart was intended for the enemy, is the fact that a part of it consisted of 90,000 buttons, marked with initials "CSA," which it is well understood stands for the words "Confederate States of America" or "Confederate States Army," the buttons being such as are used on army clothing for the three services of the army."[8]

Another report from the book *The United States versus the Schooner Stephen Hart and Her Cargo* states:

"On board were found large quantities of army and navy buttons manufactured in England by the claimants Isaacs & Co. as appears by the stamp upon them and the further impress of the initials of the same name assumed by the rebels for the country whose independence they are seeking to establish by war, CSA."[9]

Appendix L
SIC & Co Pattern 1853 Cavalry Sword and 1827/45 Cavalry Officer's Sword

The P1853 Cavalry Sword
(From the David Burt Collection)

A regulation pattern 1853 cavalry saber manufactured for SIC & Co. and exported to the Confederacy bears the unusual stamp of ISAAC & Co. on the blade. This sword is the only weapon or accoutrement to bear this stamp; all other accoutrements supplied to the Confederacy have the more familiar stamp of S. Isaacs Campbell & Co/71 Jermyn St/London. The blades were produced in Solingen, Germany, and stamped using individual dies, so to use the simpler abbreviated stamp would have saved both time and money.

The sword was described in orders of the day as "a new sword, essentially a thrusting sword."[1] It was the first sword with the tang being an extension of the blade in order to increase its strength. The first test of the sword came during the Crimean War. Although very few swords of this pattern were available when the regiments embarked, (most still carrying the 1821/22 pattern sword) during the course of the campaign, many P1853 swords were sent from the regiments left at home. By the time of the famous Charge of the Light and Heavy Brigades, half the troops carried the sword. At Balaklava, it is known that some troopers of the 11th (Prince Albert's) Hussars and 2nd (Royal North British) Dragoons were armed with this sword.[2]

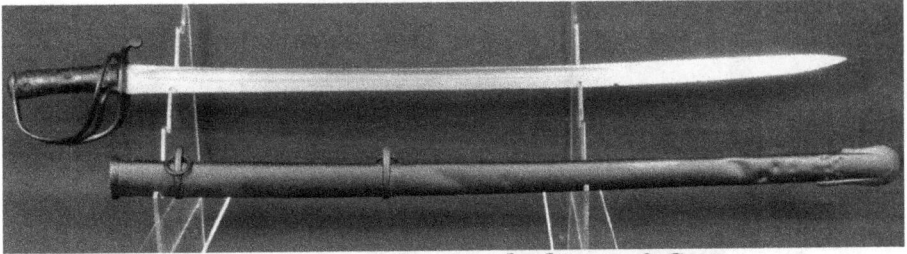

P1853 Cavalry Saber Marked Isaac & Co.
(Courtesy of Tim Prince, College Hill Arsenal)

The sword was found to have several failings, with two main disadvantages being found in the grip and hilt. The hilt offered no protection to the hand except from a direct cut and, even then, the bars were somewhat weak. As for the grip, it was leather and riveted to the blade tang and was almost rounded, which caused the sword to twist in the hand.

The strength of the blade was also put into question causing the authorities to test production samples. Weaknesses were discovered. The faults would not be remedied until 1864 when a new pattern was introduced with a sheet hilt, known as the Maltese cross, which was made with a new blade and grip.

Shortly after his arrival in England, Anderson ordered large quantities of the Pattern 53 swords. On August 3, 1861, he wrote in his diary: "Wrote to Isaac and Campbell ordering an additional number of sabers so as to number 1,000."[3] Huse purchased another "1,400 cavalry swords,"[4] costing £1,172,10s ($5,850) on March 10, 1862. These swords would be part of a shipment of 1,850 sabers, plus 992 saber belts shipped on board the steamer *Minna* – which served as a transport ship between Liverpool and Nassau – arriving in Nassau in April 1862.[5]

SIC & Co P1853 Cavalry Sword
(Courtesy of Civil War Preservations)

The above picture is a P1853 sword imported by Georgia to arm her cavalry units. Each of these swords was marked ISAAC & Co. on the blade spine and a G was stamped at the ricasso – the length of the blade just above the guard or handle. The three-branch guard is of iron, and the grip is made of two pieces of very thick checkered leather riveted to the wide blade tang.

The following excerpt is taken from the Compiled Service Record of O.C Hopkins, Captain, 1st Battalion Georgia Cavalry.

"July 10, 1862
"Camp Penn as part of 1st Battalion, Georgia Cavalry

SIC & Co. Pattern 1853 Cavalry Sword and
1827/45 Cavalry Officer's Sword

"20 English Sabers and Belts

"August 26, 1862
"17 English Sabers and Belts"[6]

The 1st Battalion Georgia Cavalry was organized with five companies during the late fall of 1861 and composed of men who had enlisted for six months service. Reorganized after the term expired, the 1st Battalion served along the Georgia coast until January 1863 when the battalion was merged into the 5th Georgia Cavalry.

The Isaac & Co. Stamp on the P1853 Cavalry Sword.
(From the David Burt Collection)

By 1864, the P53 saber was considered by the British to be an obsolete cavalry weapon, but the Confederacy continued to buy and import them right up to the end. According to the records maintained by Captain John M. Payne of the Ordnance Bureau, 34 cases of cavalry swords/sabers were shipped to Wilmington between July 17, 1863 and January 12, 1865.[7]

It is believed that approximately 5,000 of these sabers were imported into the Confederacy during the war.

The P1827/45 Cavalry Officer's Sword

The sword had an iron hilt and was officially referred to by the British as a Pattern 1827/45 Officer of Rifle Regiments Sword. The 45 in the designation indicates that the sword was manufactured with the post-1845 pattern blade.

In most respects, the sword differed little from the standard British Pattern 1822 officer's sword. However, the pattern dispensed with the folding guard feature of the 1822 sword and used iron for the hilt and guard instead of brass.

The 1827 Light Cavalry Officers Regulations stated: "The sword is to have a three bar hilt (that is a knuckle-bow and two bars outside) with a pipe back. The scabbard being of steel with a large shoe or drag."[8]

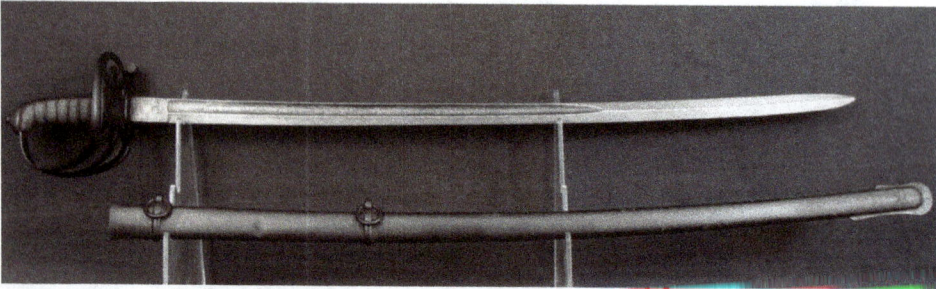

**P1827/45 Cavalry Officer's Sword with a
CSA Patriotic Etched Blade.**
(Courtesy Tim Prince, College Hill Arsenal)

Due to problems with blade breakage, a more robust blade design was officially adopted in 1845. The blade design was pioneered by Henry Wilkinson & Son during the early 1840s and was significantly stronger than the original P1822 pipe-back design. By the mid-1840s, the folding guard was no longer a common feature (due to breakage) and stronger guards became standard.

An undetermined number of the swords were produced and exported to the Confederacy. These swords were specially made for the Confederate market. Instead of the usual Rifle Regiment etching and adornments, the Confederate swords featured patriotic symbols and etchings, and were stamped with the familiar S. Isaac Campbell & Co/71 Jermyn Street/London stamp.

SIC & Co. Pattern 1853 Cavalry Sword and
1827/45 Cavalry Officer's Sword

Other features included a steel gothic hilt, with mounts of steel, a wooden grip bound with fish skin and steel wire and a steel scabbard having two bands and loose rings. It is still used today by the Royal Artillery for parade duties.

The sword in the photograph on page 158 has a Confederate themed etched blade. It has 11-1/2" etched panels on both sides of the blade. The etching shows foliate splays on both ends, with an eagle, head turned to the left, in the center. The eagle has a shield within his breast, which is engraved CSA and is surmounted by eleven stars, representing the eleven Confederate states. Another variation of this same blade etching is known to have a raised CSA in the shield, instead of an engraved one.

Halfman & Taylor of Montgomery, Alabama, were known to have imported this exact variant of officer's sword with the obverse ricasso etched with their name and address instead of the SIC & Co. name and address.

The P1827/45 Officer's Sword

The SIC & Co. Etching
(Courtesy Tim Prince College Hill Arsenal)

Appendix L

The reverse ricasso of the sword is etched in five lines: S. Isaacs Campbell & Co/71 Jermyn St/London. The obverse ricasso is etched with a six-pointed star surrounding a brass disc within the ricasso.[9]

Proof of purchase by the Confederacy comes from an invoice in the McRae Papers, which clearly refers to the purchase of officer's swords. The invoice from SIC & Co. lists: "59 Officer's Swords – 19s.6d" ($95).[10] According to the invoice, the case was to be marked OS/59 in a rhomboid.

Like so many of the swords and saber bayonets imported by the Confederacy through SIC & Co., this sword is of Solingen origin. The 1827 pattern with its three bar hilt was still a regulation cavalry officer's sword when the war broke out in 1861. It would be replaced in 1896 by the Heavy Cavalry sword.

Appendix M
Other War Materials Supplied By
SIC & Co.

Blankets, Cloth and Greatcoats

When the *Stephen Hart* was captured, she carried a typical cargo of war materiel for the Confederacy. This is the inventory of the diverse range of arms, equipment and other stores she carried.

* 5,740 long Enfield Rifles with triangular bayonets
* 1,260 short Enfield Rifles with saber bayonets
* 660 Enfield Carbines with saber bayonets
* 2,640 British Muskets with triangular bayonets
* 320 Brunswick Rifles
* 6,800 gray blankets
* 1,750 white blankets
* 4 Blakely 2-3/4 inch bore rifled cannon (six pounders) with 1,008 shells for the same, loaded and capped
* 100,000 percussion caps
* 100,000 Brunswick rifle cartridges
* 420,000 minie rifle cartridges
* 2,160 cartridge boxes
* 4,095 knapsacks
* 4,000 ball bags and belts
* 1,540 yards of gray cloth for uniforms
* 11,453 yards of steel gray mixed cloth for uniforms
* 625 gross of buttons for uniforms, marked CSA
* 15,432 pairs of stockings
* 2,000 pairs of brogan shoes
* 592 pairs of russet shoes, Blucher Pattern
* 762 pairs of black shoes, Blucher Pattern
* 2,220 waterproof covers for mess tins
* 17 cases of trimmings and 3 bales for army clothes and uniforms, consisting of linings, cord, braid, lace and thread

* ★ 109 yards scarlet cloth for uniforms
* ★ Surgeon's equipment
* ★ 7,500 yards of white twilled flannel for lining army overcoats
* ★ 2,250 yards of brown holland for the same purpose
* ✶ Appurtenances for small arms, gun slings, medicine, lint and bayonet scabbards[1]

Blankets

Blankets were items that were sorely needed right from the outset of the war and were one of the first things on Huse's shopping list.

An October 11, 1861 SIC & Co. invoice for shipment on the *Fingal* mentions "970 blankets & 5,240 blankets"[2] and were part of the 9,982 yards of blankets that were on board. The *Fingal* arrived in Savannah on November 15, 1861.

The *Gertrude's* cargo included: "3,960 pairs of gray army blankets & 335 pairs of white blankets."[3] The *Springbok* carried "540 pairs of 'gray army blankets,' like those used in the army of the United States, and 24 pairs of 'white blankets.'"[4]

On February 3, 1863, Gorgas reported the safe arrival of 62,025 blankets.

Cloth

On December 5, 1862, Gorgas wrote Seddon:

> "In addition to ordnance stores, using rare forecast, he (Huse) has purchased and shipped large supplies of clothing, blankets, cloth and shoes for the Quartermaster's Department without special orders to do so."[5]

By February 1863, Gorgas had reported the safe arrival of 78,520 yards of cloth from England. Most, if not all of this cloth, had been purchased by Huse. SIC & Co.'s November 1861 invoice includes "gray army cloth."[6]

Other invoices reveal:

* ☆ December 12, 1861: Gray Army Cloth
* ☆ December 14, 1861: 7,250 yards Blue Gray Army Cloth
* ☆ December 16, 1861: 23,954 yards of Oxford Gray Army Cloth
* ☆ December 24, 1861: Blue Army Cloth
* ☆ July 24, 1862: Blue Gray Army Cloth[7]

Approximately 24,145 of the 78,520 yards of imported cloth reported by Gorgas can be accounted for in surviving invoices from SIC & Co. and in correspondence between Huse and Gorgas. While 9,645 yards of this was blue gray and oxford gray cloth - as accounted for in the SIC & Co. invoices – the biggest amount was light blue cloth – used for trousers – which totaled 14,500 yards.

SIC & Co. purchased most of their cloth from Yorkshire woolen mills. Two letters from a Halifax woolen mill mentions the proposed purchase of "light gray mixture and blue gray mixture."[8]

Greatcoats

British regulations for greatcoats stated: "Greatcoat, made of gray kersey in four different sizes, weighing 6 lbs. to 6 lbs. 4 oz., and are to last 4 years."[9]

A greatcoat consisted of a laydown collar, single breasted with a six button front and two buttons on the cape. It also featured a button and button hole on the hem of the coat to allow the skirt to be buttoned back if desired. Most of the imported greatcoats would have been supplied in the Confederate regulation uniform color of blue gray kersey, although an advert from an Augusta auction sale stated: "100 military gray overcoats, English regulation in oxford gray cloth."[10]

The *Stephen Hart* manifest includes "7,500 yards of white twilled flannel and brown holland"[11] for lining army overcoats. This is in addition to the 8,675 British greatcoats successfully imported up until February 1863.

Huse purchased "450 artillery and 200 infantry greatcoats"[12] on December 24, 1861. On January 22, 1862, another "845 infantry and 671 artillery greatcoats" worth £1,954 ($9,770)[13] were invoiced and shipped on board the *Economist*.

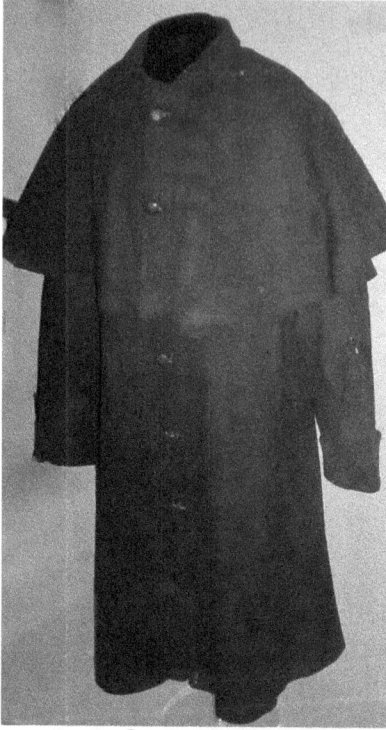

An Infantry Greatcoat
(Courtesy of Kevin Dally and the Texas State History Museum)

The only known surviving Confederate-used English imported greatcoat is now on display at the Texas State History Museum in Dallas. The museum describes it as: "A Confederate overcoat made of cadet gray wool and worn by an unidentified soldier from Virginia."[14] It is also made of what appears to be a lesser quality kersey and features the button and button hole at the bottom of the skirt to enable it to be buttoned back.

APPENDIX N
OBITUARIES

Samuel Isaac

We regret to announce the death of Major Samuel Isaac, the Lesseps* of the Mersey Tunnel, who expired on Monday afternoon at his residence in Warrington Crescent, Maida Vale, at the age of 74. Major Isaac was born in Chatham, came to London as a young man, and carried on a large business as an army contractor in Jermyn Street, under the firm of Isaac Campbell & Co. His brother, Mr. Saul Isaac, JP, formerly Member of Parliament for Nottingham, being another member of the partnership.

The firm was, during the Confederate War in America, the largest European supporter of the Southern states and their ships laden with military stores and freighted home with cotton, were the most enterprising of the blockade runners.

The late Major Isaac's eldest son, Mr. Henry Isaac, who died in Nassau, West Indies, during the war, had much to do with this branch of the work.

Major Isaac's military rank was conferred on him in connection with his services in raising a regiment of volunteers from among the workmen at his own factory in Northampton. Messrs Isaac Campbell & Co. was naturally large holders of Confederate bonds. The commercial house fell shortly after the fall of the Confederacy, and Major Isaac's enormous mansion at Kensington, tenanted after him for a season by the Begum of Oudh, long stood vacant.

He was not, however, the man to be daunted by failure. After a time, he acquired the rights of the promoters of the Mersey Railway – a project which had obtained the sanction of Parliament, but had remained dormant owing to the disinclination of capitalists to venture on the heroic task of tunneling the bed of the Mersey. With unfailing courage and persistence, Major Isaac pushed the scheme into practical development. He himself undertook to make the tunnel, letting the works to Messrs Waddell and seeking the invaluable assistance as engineers from Brunlees and Sir Douglas

Fox. Fresh powers were obtained from Parliament, money was raised in bonds and shares and the tunnel was duly opened under the auspices of the Prince of Wales.

His first wife was Miss Symonds of Dover, by whom he had three children. He married secondly, Emma, daughter of the late Stephen Hart, of Haydon Square, London, who survives him, as does their daughter Mrs. Arnold Crombach, with other issue. The funeral was solemnized yesterday (Thursday) at the Willesden Cemetery of the United Synagogue.[1]

*Ferdinand De Lesseps was the French designer of the Suez Canal

Saul Isaac

Mr. Saul Isaac, MP for Nottingham from 1874 to 1880, died on Tuesday at 109 Greencroft Gardens, South Hampstead, and he is to be buried today at Willesden, the funeral leaving the house at 12 o'clock.

He had a chequered and romantic career. He was born eighty years ago, and his early years were passed at Chatham, where his family was engaged in the furniture business. The experience of military requirements gained in this important garrison enabled elder brother, the late Major Samuel Isaac, and himself, to become, at a later period, the principal army contractors and European agents of the Confederate government in the struggle between Northern and Southern States of the great American Republic.

Their firm at this date was Campbell Isaac & Co., of Jermyn Street, and their business was to run the blockade which the more powerful fleet of the Federals maintained against the Confederates. Mr. Saul Isaac's nephew, Henry, a daring and handsome young man, went out to Nassau, West Indies, to superintend these operations, and died there of yellow fever. The liabilities of the Confederates were paid in bonds, and when the defeat of the South became definitive, Mr. Isaac's firm was the largest holder of these bonds. With the fall of the Confederation, its principal European commercial supporters fell also.

A remarkable fact was the subsequent financial recovery of both partners. They had married sisters, the misses Hart, ladies who had inherited substantial separate estate, and after resigning their mercantile assets and their well appointed private establishments to their creditors, the brothers had still from their wives' means the opportunity of starting again.

The elder brother became the maker of the Mersey tunnel, and thus acquired a second fortune, though ultimately he left his affairs considerably involved. The younger brother, the subject of this memoir, became the owner of Clifton collieries at Nottingham. Shortly afterwards, he was elected for the important borough (now city) near which his collieries were situated; and he was bidden to Windsor by the Sovereign.

In those days, Jewish emancipation being comparatively recent, the Jewish members were all Liberals, and Mr. Isaac was the first Jew to take a seat on the Conservative side, the first Jewish Member of Parliament to support Lord Beaconsfield, the first Jew elected a member of the Carlton Club.

He was a handsome young man of good presence, but speculative and unstable in his business affairs, and his second period of brilliance proved as fleeting as the first. He fell into difficulties from which he is believed never to have recovered.

Mr. Isaac was married twice and leaves issue. Mr. Isaac continued to take interest in Jewish matters and write to us as recently as August 13, 1902, in reference to what he was good enough to call "our very able article" on "some events in the days of the King."[2]

Caleb Huse

Colonel Caleb Huse Dead
Northerner was South's Purchasing Agent during Civil War
Highland Falls, NY, March 12.

Colonel Caleb Huse, aged 75 years, died suddenly at his home here today, following a surgical operation. Colonel Huse was graduated from the United States Military Academy in 1851, and for many years was an instructor at West Point.

He resigned from the Union Army in 1861, and subsequently was commissioned by Jefferson Davis as colonel and sent to Europe as purchasing agent for the Confederate Army. He remained abroad in that capacity until the end of hostilities. For a year prior to the war of

167

the rebellion, he was Superintendent and Commandant of Cadets at the University of Alabama.

Colonel Huse was born at Newburyport, Massachusetts. He is survived by his wife, three sons and five daughters. One son, Harry J. P. Huse, is a professor of mathematics at the Naval Academy.[3]

Part 2
Peter Tait & Co., Limerick

Sir Peter Tait
(Courtesy Limerick City Council Archives)

Sir Peter Tait is a name familiar with all Civil War re-enactors, living historians and collectors around the globe. A name famous for uniforms he provided to the Confederate States of America and the jacket no authentic re-enactor can do without for his late war impression.

CHAPTER 17
A SUIT OF BLUE

During the last two years of the war, there were numerous reports of soldiers in the Army of Northern Virginia who appeared to be wearing blue-gray jackets produced of imported English broadcloth or woolen kersey. One of the first accounts comes from Ted Barclay of the 4th Virginia Infantry, Company I, of the famed Stonewall Brigade, who recalled in late May, 1863: "...as I was getting tolerably ragged, the brigade secured a supply of English clothes. So as I was one of the needy ones, I am rigged in a splendid suit of blue."[1]

Private Augustus Dickert, 3rd South Carolina Volunteers, Company H (Kershaw's Brigade, Longstreet's Corps) noted that their uniforms consisted of "a dark blue round jacket closely fitting with light blue trousers; it closely resembled those worn by the enemy, the only difference being the cut of the garments – the Federals wearing a loose blouse instead of a jacket."[2] By late summer 1863, entire brigades were outfitted with new jackets made from English kersey wool. An Illinois infantryman described the Confederates as: "better dressed than we are, their uniforms being apparently new... The Carolinians' uniforms are a bluish gray...with sky blue pants."[3]

At the Battle of Chickamauga, Longstreet's Corps caused considerable confusion in the Confederate ranks. Captain Frank T. Ryan of the 1st Arkansas Mounted Rifles wrote:

> "When we learned of the dangerous situation ourselves, we halted. In the meantime, the troops in our rear were coming steadily towards us...some insisted they were Longstreet's men and therefore our friends, others said they could distinguish them plainly and they were Federals. How such a difference of opinion could arise was owing to how Longstreet's men were uniformed. They wore light blue pants, gray jackets and regular soldiers caps."[4]

Chapter Seventeen

Even General Ulysses S. Grant experienced an embarrassing encounter when he arrived in Chattanooga weeks later.

> *"General Longstreet's corps was stationed there at the time and wore blue of a little different shade from our uniform. Seeing a soldier in blue on this log, I rode up to him, commenced conversing with him and asked whose corps he belonged to. He was very polite and, touching his hat to me, said he belonged to General Longstreet's corps. I asked him a few questions – but not with a view of gaining any particular information – all of which he answered, and I rode off."*[5]

In November 1863, at the Battle of Kelly's Ford, a Union soldier recalled that their Confederate prisoners were better outfitted than the Federals were, "with uniforms of English manufacture, much darker (blue) than the United States uniform, and this furnished conclusive evidence of successful blockade running."[6]

★ ★ ★

The official Confederate uniform regulations published on June 6, 1861, called for a "double breasted tunic of gray cloth, known as cadet gray."[7] Huse, with the aid of SIC & Co., was able to locate similar dark blue-gray kersey produced in a woolen mill in Halifax, Yorkshire. The kersey was used to produce British military greatcoats and trousers worn by staff sergeants, sergeants and the rank and file of Guards regiments. In a letter to Samuel Isaac, the mill owner described the material as a "blue gray mixture."[8] SIC & Co. invoices note the supply of this cloth early in the war. The December 14, 1861 invoice states that 7,250 yards of "blue-gray army cloth"[9] being shipped on the steamers *Economist* and *Southwick*. Other invoices mention gray army cloth and blue gray cloth being purchased widely from November 1861 to July 1862.[9] By the end of December 1862, Huse had bought and shipped 78,520 yards of cloth.

By the spring of 1863, the lack of cloth for uniforms became a serious concern for the Quartermaster's Department and, particularly, the Richmond Depot. Major Richard Waller, the depot's superintendent, summed up the situation in a letter to Myers:

"The greatest auxiliary to this depot and others will be to sustain Major Ferguson in England with funds. He has made purchases already to a considerable extent, as a small portion of which has arrived here. He has purchased 6/4 woolen cloth at 4 pounds Sterling, say 96 cents, which, adding exchange and transportation, I estimate to cost here $2.50 to $2.75 per yard. The exchange with which these purchases were made was bought by him at about 125% premium.

"For cloths, of an inferior quality to that purchased by Maj. Ferguson, I have been forced to pay $12.00 per yard, within the last six months and, at this time, from $7 to $8.50 per yard. And I do not think out of the present supply of wool, I can have goods manufactured of the quality made by the mills in this vicinity, at less than present prices.

"I do not think that the average supply of wool in the Confederacy is adequate to the wants of the government and the people, neither have we woolen mills enough. And I, therefore, respectfully and earnestly recommend that arrangements be made to procure a supply of material for clothing in England. This is a question of great importance."[10]

Alexander Lawton (having replaced Myers as Quartermaster General) ordered Ferguson to buy as much cloth as possible. Ferguson had secured the products of many of the "Lancastershire (sic) and Yorkshire manufacturers"[11] to supply the blue-gray woolens needed for the uniforms. By mid-October, Ferguson was able to report to Lawton the successful purchase of over one million yards of cloth from these woolen mills. Ferguson wrote, "Having been an importer for several years, I was not unfamiliar with the various kinds of the goods ordered and well acquainted with the proper locality where they could be procured on the best terms."[12]

Records indicate that the Quartermaster's Department began to receive large quantities of the blue gray cloth from late 1863 onwards. But by April 1864, Lawton was still troubled by the lack of domestic cloth for uniforms. He wrote McRae: "The limited quantity of wool in this region of country compels me to look abroad for material for clothing."[13]

On June 10, 1864, the Richmond Depot received 4,574 yards of cloth shipped by Ferguson followed by an additional 4,983 yards on June 13. Three days later, 2,983 more yards were brought in, making a grand total of 12,540 yards of blue gray cloth. The cloth was described by the depot as: "English gray cloth, English blue cloth and English gray-privates."[14]

The Richmond clothing manufactory had become largely dependent on the blue gray kersey sent by Ferguson. Lawton wrote Ferguson: "In other respects, you are fully advised as to its winter wants, and you will continue to purchase and ship, as heretofore, of the best quality of gray cloths, with a fair proportion of trimming."[15] As a result, supplies of the imported blue-gray cloth stayed fairly constant from the summer of 1864 until the final months of the war.

From November 5 to December 4, 1864, 48 bales of cloth, equaling 27,648 yards, were received into Wilmington and Charleston from England and warehouses in Nassau and Bermuda.[16]

★ ★ ★

When viewed from a distance of more than a few yards, the surviving jackets made from this cloth exhibit a dark bluish gray tint. When examined closely, the material is variable in tone. When viewed under magnification, the wool is revealed to be a combination of both dark blue and light to medium gray fibers. The ratio of blue to gray fibers is roughly 60:40.

The cloth was apparently dyed in two distinct colors and then carded together before spinning into yarn. As synthetic dyes first came into use in England around the time of the war, it is likely they were used in making this particular cloth. The use of good, fast and deep dyes helps explain why surviving jackets show very little signs of fading. The surviving examples lend excellent insight into the authentic coloration of the garments.

Thomas Arliskas, author of *Cadet Gray and Butternut Brown*, conducted research into numerous period accounts concerning what Confederates soldiers at various times were wearing. He noted:

"...beginning in mid-1863, there were more uniformed troops in the ranks. However, there was still a quantity of civilian and homemade uniforms along with Quartermaster-issue, as well as government purchased homemade garments, all found in the same company and regiment. In the autumn of 1863, whole brigades were finally uniformed the same."[17]

Surviving records indicate that before departing for Chickamauga, Longstreet's Corps was issued Richmond Depot (type II) jackets of English blue-gray wool. These jackets were made by various Confederate clothing bureaus out of the imported English gray cloth.

Due to the capture of several blockade runners of the Crenshaw/Collie line in early 1864 and Ferguson's inability to find new suppliers in the already swamped British market, a new source of uniforms had to be found.

By the end of 1864, a new type of ready-made uniform appeared in the ranks of the Army of Northern Virginia, Army of Tennessee as well as troops in Alabama and the Trans-Mississippi Department of Texas, Louisiana and Arkansas. This uniform consisted of a close fitting jacket and matching trousers, with the jackets having various trims and made out the same blue gray cloth as the domestically produced uniforms.

These new ready to wear uniforms were produced by the Irish company of Peter Tait & Co. of Limerick, who was a major supplier to the British Army.

Although made from the same color blue gray material, the Tait contract jackets differed from domestically produced jackets in the following ways:

★ Made of fine grade kersey or broadcloth.
★ Five piece body (no back center seam).
★ Lined in domestically produced Irish linen.
★ Eight button front with variously designed script I (Infantry) or A (Artillery) buttons. These buttons, unique to the Tait jackets, had a recessed center and a floating shank.
★ Curved cut front collar.

* ★ Double line of off-white or brown machine topstitching on button side.
* ★ White or brown buttonhole thread.
* ★ Inside left vertical pocket.
* ★ Machine sewn, except buttonholes, and double line of stitching on the button side of the front closure.
* ★ Marked on the lining in British Army chest and waist sizes.

These handsome looking jackets later became known simply as the Tait jacket.

CHAPTER 18

PROSPERITY TO THE TRADE OF LIMERICK

Peter Tait was born in Lerwick, Scotland,[1] in 1828. He was the son of Thomas and Margaret Tait, who owned a greengrocer's shop.

In 1844, Thomas secured an apprenticeship for his son, who was now sixteen, with the Scottish firm Cumine, Mitchell and Co., a Limerick drapery firm. During the Great Potato Famine (1845-1849), Tait was released from his apprenticeship. To keep his job, he offered his services for room and board, but to no avail.

Tait noticed that many sailors visiting Limerick were in need of shirts, so he sold ready-made haberdashery items from a basket on the docks. His hawking skills produced prosperity during the lean famine years. By 1850, he had rented number 4 Bedford Row in Limerick and employed a woman, who, by using a sewing machine, was able to make all the shirts he needed.

In 1856, Tait married Rose Abraham, daughter of William Abraham, a local nurseryman, and his wife, Elizabeth, of Fort Prospect, Limerick. Tait took out an 100 year lease on South Hill House in Rabane South in the County of Limerick. On the porch, paying homage to leaner days, sat the basket Tait had used to hawk shirts on the docks.

Tait saw great opportunities in William Thomas' lock stitching sewing machine. He purchased one and took it apart to familiarize himself with the inner workings. He spent £4,000 ($20,000) to purchase Thomas sewing machines for Bedford Row and a new workhouse on Edward Street.

Traditional clothing manufacturers employed seamstresses, who worked at homes. Tait, on the other hand, placed all his workers under the same roof. Each worker completed one task and passed the garment down the line. By the time the garment reached the last operator, it only required the finishing touches. Tait had invented the world's first production line, which reduced the cost of making a garment.

Tait secured two-year contracts to supply uniforms for British line and rifle regiments serving in India. He was called to testify in front of the Royal Commission and informed the commissioners that his company had supplied 120,000 uniforms between 1855 and 1858. He stated that his factory could provide 250,000 uniforms a year, or 10,000 a week. If he had the orders, he said, he could clothe the whole British Army.[2]

In 1850, George Cannock and John Arnott had purchased Cumine, Mitchell and Co., changing the name to Arnott, Cannock and Co. In 1858, Tait used his savings and an inheritance he received from his Uncle Peter to buy Arnott's share. The firm was renamed Cannock, Tait and Co. and became the official uniform supplier to the British Army. Tait left the daily operation to his partners and focused on the business on Bedford Row.

In the pockets of his uniforms, Tait slipped a small token stamped with the slogan: "prosperity to the trade of Limerick." He took out a 999 year lease on an old auxiliary workhouse and transformed the property into the most modern clothing factory in Europe. The new factory covered three acres. The main machine room was 300' x 100' and was divided into three bays. There were five long tables with gangways in between. Over 200 sewing machinists, basters, buttonhole makers, finishers and pressers were employed – a total of between 700 to 900 workers. The majority of the workers were women. Tait paid good wages and looked after his employees. In return, he was well thought of and respected.

A journalist with the *London Times* newspaper visited the factory on August 3, 1867. He reported:

> "The long workroom contains 150 sewing machines which employ 500 work girls. All goes by steam; so that the doctors who cry out about the evil effects of the machines in the London slop-shops can now point out a remedy. The ventilation is perfect; and the neatness of the girls contrast strongly with what one sees in the Yorkshire and Lancashire cotton mills. Of course, in other rooms, there are cutting machines which go through 24 thicknesses of cloth as easily as you would go through cheese and pressing irons heated inside with gas and all the most modern adjuncts of a great clothing

establishment. But the main point is these girls, who would else have been picking up a wretched livelihood by making a little lace and hawking it about the streets, get here from 8 to 10 shillings a week."[3]

The Army Clothing Factory, Limerick. The blocked up red archway provided carriage access to the factory. Confederate uniforms may well have passed through it on their way to the war.
(Courtesy of the Limerick City Council Archives)

CHAPTER 19
A CONTRACT WITH THE
CONFEDERATE GOVERNMENT

The man who would play a vital part in the story of Peter Tait & Co. supplying uniforms to the Confederacy was Peter's elder brother, James. James Linklater Tait was born in Lerwick in 1824. He established a drapery business in Aberdeen, Scotland, and, in 1851, opened a second branch in Kirkwall on the Orkney Islands. In 1854, James partnered with Balfour Logie. The name of the company was changed to Tait & Logie, but the business was forced to close in 1860.

In 1861, James and younger brother, Robert, established a London subsidiary of the main factory in Limerick – Tait Bros & Co. The site made uniforms for smaller orders, bought and stored buttons and leather goods for the Limerick factory and was a distribution site for uniforms shipped from Ireland. The London factory provided Confederate officers' uniforms. Excavations done on battlefields in the South have found officer buttons bearing the CS Eagle design, size 15/16", back marked RT Tait & Co/Essex Street Strand/London. After James left the firm to serve as Peter Tait & Co.'s agent in Richmond, Robert changed the name of the company to Robt Thos Tait & Co.

On December 15, 1863, James Tait wrote Seddon:

> "Sir,
> "I had the honor to submit to you on the 9th inst an offer to supply Army cloths and other necessaries for Confederate Military Service.
> "In doing so, I stated my position in the British Military Service as Captain and sub inspector of stores and the peculiar facilities I had from my position for supplying cheaply and superior quality of any stores of that description.
> "As directed by you, I waited upon the Quartermaster General with whom, and one of his subordinate officers,

I had a personal interview on the subject. I failed, however, to discover any means within the reach of this department to test the prices and qualities of military stores supplied, nor did I ascertain that there were any gentlemen sufficiently conversant with the business whose knowledge and experience would be of value to the government in this respect.

"The advantages that from my position I can offer in contracting for supplies are fourfold.

"1st. The careful inspection and testing of all articles supplied as to durability and color by the tests in use in the military store department of Great Britain.

"2nd. The purchases of stores from the most eminent contractors in their respective branches (to whom I am personally known) at the same prices as are paid by the British government.

"3rd. A knowledge of the lowest contract prices and a thorough familiarity with every trick and detail of the trade by which I can be guided in the purchase of supplies.

"4th. A rate of remuneration which will be found lower than is paid to any other contractor or agent.

"It has been a matter of notoriety in contracting circles in England that the Confederate government has been in second or third rate hands in the matter of supplies. Goods rejected as unfit for other services have been purchased on Confederate government accounts and prices have been charged for these inferior articles vastly in excess of the value of sound and serviceable stores.

"There can be no question that apart from the actual loss sustained by such transactions, a further and severer loss is experienced in the want of confidence produced in the minds of the respectable mercantile and moneyed classes.

"Seeing the trade in doubtful hands and apparently carried on with an utter disregard of prices or qualities, they are disposed to regard its transactions in the category of gambling ventures and fight shy of it *accordingly and nothing would seem more effectually to end this state of things than the restoration of the business of safe, honorable and legitimate channels.*

"The ability to deal with manufacturers of the first standing would react beneficially upon the bonds and other securities of this government in England and thus the purchasing and selling Departments would find their facilities for both largely increased.

"Since my arrival in this country, I have been with your armies in the field and have been shocked to perceive the disgraceful cloth which has been palmed off upon your English agents and which is served out to your soldiers who are called upon to face, in the field, the inclemencies of a Virginia winter. I think it is doubtful whether one tenth of the cloth that reaches the Confederate States is genuine broadcloth, by far the larger portion consisting of shoddy cloth,[1] an article worth about one fifth the price your agents pay for it in England and not more than one tenth of the price it costs the Confederate government when paid for here.

"And I may be permitted to mention here that there is no trade in England so beset with snares and pitfalls as the cloth trade. It is well known that there are agents of these unprincipled shoddy houses lying in wait at Liverpool and elsewhere for the employees of foreign governments who are usually utterly without experience as to the qualities of cloth and who are obnoxious to all the private influences which these agents know so well to apply.

"From a sincere desire to assist the Confederate people in their mighty struggle and the conviction that, to an extent probably beyond your own conception, the Confederate government is injured and robbed and the efficiency of its soldiers imparred, (sic) I offer to place at their disposal the same facilities for undertaking contracts and the same checks upon agents and contractors with which long experience has armed the British government.

"In conclusion, I can truly say that this offer is volunteered solely because I am aware to what extent your government and people stand in need of protection. It is my hope that I can be of material assistance to them in this matter, and that it may lead to future transactions in regard to the permanent supply of clothing and other stores for the Confederate standing army which will be advantageous to your government.

"I do not wish to leave the Confederate States without at least making available for their government the experience which my official position has placed in my power and without bringing to the notice of the government the fiends, to which I am conscious, they are at present a prey. It is for them to decide whether to avail themselves of an opportunity which accident enables me to offer and which I believe will prove of eminent service to them.

"It would not require more than three months from 1st January 1864 (If my departure takes place hence immediately) to have any amount of army clothing, blankets, boots & shoes, etc. shipped from British ports to the islands.

"I have the honor to be sir your obedient servant,
"James L. Tait

"ENCLOSURE:
"To be ready for shipment in 3 months from Jan 1st 1864
"50,000 caps ready cut (Grey Cloth with Peak 1/6 ea)
"50,000 Greatcoats of Stout Grey Cloth cut ready for sewing (Fast Dye) 12/ ea
"50,000 Suits of Stout Grey Cloth consisting of jackets & trousers ready for sewing 16/ per suit
"50,000 Strong Grey Flannel shirts ready for sewing 4/ ea
"50,000 pairs Army Blankets 5/6 ea
"10,000 pairs Army Boots (Blucher) 10/6 ea
"100,000 pairs Woolen Stockings 1/ ea
"50,000 Haversacks 1/6 ea
"Total Amount: £158,475 Sterling. ($892,375)

"The clothing to be cut according to the size rolls in use in the British service to fit men from 5 feet 6 inches to 6 feet. The quality of all the supplies would be those in use in the British Army and subjected to the same rigid inspection."[2]

Lawton confirmed the order in a letter to Tait on December 19, 1863. He told Tait to pass the letter on to McRae in London for ratification. Nothing seemed to come of this arrangement. In June 1864, Tait reduced the order to clothing and shoes. Lawton wrote McRae:

"Mr. (James L.) Tait has recently renewed his proposition to furnish supplies to this dept. He offers to provide, within four months after his arrival in England, 100,000 suits of clothing, each suit to consist of a cap, jacket, trousers and shoes corresponding in quality with samples delivered here and deliver the same free on board at Limerick at 30 shillings per suit. I have accepted this offer to the extent of 50,000 suits, subject to your ratification and with the understanding that payment is to be made by you in the bonds of the Confederate States to be delivered to Mr. Tait at such valuation may be agreed upon between you. The caps can be omitted at a valuation of 3 shillings each."[3]

Once back in England, Tait began negotiations with Alexander Collie, McRae and Ferguson. McRae wrote Seddon on July 4, 1864:

"I have made a contract with Alex Collie & Co. for £150,000 clothing and quartermaster's supplies to be purchased by Major J.B. Ferguson, and for £50,000 ordnance and medical supplies to be purchased by Major C. Huse.

"...In a letter to the QM General of 7th May, I advised him that I should make a contract with Mr. Collie for his department for carrying out that made by Mr. Tait at Richmond and sent to me for confirmation. Mr. Tait has

arranged with Mr. Collie and Major Ferguson to furnish £50,000 ($250,000) worth of ready-made clothing at prices somewhat lower than those named in the contract drawn up at Richmond, and he waives the 5 percent, as Mr. Collie pays him cash for the goods on delivery."[4]

This above contract supplied £150,000 ($750,000) worth of Quartermaster's stores to the Confederate government.

CHAPTER 20
STATE CONTRACTS

While waiting for confirmation of his contract with the Confederate government, in June 1864, James Tait traveled to Montgomery, Alabama, where he negotiated a separate contract with Duff C. Green, Alabama Quartermaster General to supply uniforms, shoes and overcoats for the militia. The supplies were to be paid for by 698,000 pounds of cotton, which Tait would ship back to England to sell to the cotton-starved Lancashire mills. Collie advised against the idea, but Peter Tait agreed to the order, thinking the arrangement was fiscally sound.

The *Columbus Daily Enquirer* (Georgia) published the following story:

> *"October 22, 1864*
>
> *"GOOD NEWS FOR ALABAMA SOLDIERS*
>
> *"Four months ago, a contract was entered into between the State of Alabama on the part of the Quartermaster General and the firm of Peter Tait and Co., Limerick, Ireland, through Major J.L. Tait, of the British Army, for a large quantity of military clothing for the Alabama soldiers. Quartermaster General Green stipulated that a large portion of the goods should be furnished partially cut, with the necessary trimmings, thus affording employment to the seamstresses and tailors of our home factories. Some thousands of these uniforms, we are glad to be able to announce, have safely arrived in the Confederacy, and the residue of the order is hourly expected. The outfit consists of jacket, pants, shoes and overcoat all made of the most substantial material – the cloth being exactly the same as used in the British Army.*

"The house of Peter Tait and Co. in Limerick, is one of the most extensive in Great Britain, and these enterprising factories furnish a greater portion of the outfit to the British Army than any other in the realm and give employment to two thousand operatives. Their own vessels run into the Confederate ports, and they have filled frequent contracts with the government. Their contract with the State of Alabama has been faithfully and promptly fulfilled, thanks to the energy and tact of their agent and representative, Major Tait.

"Some of the goods for our State troops is (sic) already made up into uniforms. A specimen overcoat, which we have seen, exhibits the superiority of the material for durability and comfort and the excellence of the make-up.

"When Alabama's soldiers are comfortably clad in the ample folds of a cape and skirt of one of these overcoats they will be as handsomely and as comfortably uniformed as any soldier in the world. Several thousand of these uniforms are already here. The rest of the order will be here in a few days. Montgomery Mail."[1]

In March 1863, a new pattern greatcoat for the British Army, made from "two qualities of gray kersey"[2] was created. These greatcoats, as well as uniforms, were sent in kit form to Alabama. On October 8, the first order arrived at Wilmington on the *Condor*. The *Condor* was one of the "four large and powerful new steamers"[3] provided by Collie for his contract with McRae.

S.M. Eaton, Union Chief Signal Officer, Military Division of West Mississippi, wrote to Lieutenant Colonel Christensen, Assistant Adjutant General, about the new uniforms.

"I have the honor to make the following report of information received at this office this 23rd day of January 1865. William Ross, a deserter from the 2nd Alabama Battalion, left Mobile January 14, 1865. He states that General Thomas' command consist(ing) of Colonel Rice's brigade (three regiments) state reserves, 1,200; brigade (two regiments) state reserves 1,500;

total state reserves, 2,700. These troops are pretty well armed, well clothed with a late importation of gray suits from England."[4]

According to Ross, more than 5,500 Alabama soldiers from General Bryan Morel Thomas' brigade were clothed in the Tait style uniforms.

Confederate nurse Kate Cummings wrote in her diary about the finely dressed Alabama soldiers.

"Mobile never was as gay as it is at the present; not a night passes but some large ball or party is given. Same old excuse; that they are for the benefit of the soldiers, and indeed the soldiers seem to enjoy them. The city is filled with the veterans of many battles. I have attended several of these parties, and, at them, the gray jackets were conspicuous. A few were in citizens clothes, but it was because they had lost their uniforms. The Alabama troops were dressed so fine that we hardly recognized them. A large steamer, laden with clothes, ran the blockade lately from Limerick, Ireland."[5]

★ ★ ★

Private Henry Pillans' (62nd Alabama) jacket is the lone surviving Alabama contract jacket. It was a typical Tait jacket in appearance. It was made of the fine blue gray cloth with no back center seam. The jacket had a solid blue collar trim and shoulder straps.

The jacket differed from a factory machine made jacket in several ways. The Alabama jacket was originally made with a five button front, instead of the eight button front normally associated with the Tait factory. Four buttons and buttonholes were added later. The original buttonholes were sewn using white thread, and the additional buttonholes were sewn with brown thread. It is probable that the jackets provided to Alabama did not include the smaller 3/4" buttons with the recessed center used by the Tait factory. The Pillans jacket had normal sized brass buttons provided by the state. The bottom edges of the jacket were rounded instead of straight cut. The jacket also lacked the double row of machine stitching on the button side.

Pillans' Contract Jacket
(Courtesy of the Mobile Museum of History)

★ ★ ★

The Trans-Mississippi Department

During the latter part of 1862, the Trans-Mississippi Department purchased cotton in order to buy much needed quartermaster and ordnance goods from England. The supply situation had become critical after Grant captured Vicksburg in July 1863 and severed the Confederacy in two.

By autumn, the Ordnance, Quartermaster's, Commissary and Medical Bureaus' agents were all vying for the available cotton in the region. Lieutenant General Edmund Kirby Smith, in command of the Department, created a new Cotton Bureau to consolidate the agents' efforts. Kirby Smith chose Lieutenant Colonel W.A. Broadwell to head the new bureau.

> "HDQRS. Trans-Mississippi Department,
> "No. 35
> "Shreveport, La., August 3, 1863.
>
> "Lieutenant Colonel W. A. Broadwell is announced as chief of the Cotton Bureau of the Trans-Mississippi Department. All government agents for the purchase, collection, or other disposition of government cotton are directed to report to and receive their instructions from Lieutenant Colonel Broadwell.
>
> "By command of Lieutenant General E. Kirby Smith:
> "S. S. Anderson, Assistant Adjutant-General"[6]

Upon his appointment, Broadwell undertook a thorough survey of the cotton situation and designed an organization to purchase and house the cotton under one roof. Kirby Smith selected Major Joseph Minter as the Bureau's agent abroad.

In March 1864, Broadwell ordered the assembly of 2,500 bales of cotton to pay for the departments' stores. The assembly progressed slowly. It was summer before Minter could finally sail for England. He arrived on July 29 and met with McRae and Ferguson to procure ordnance and quartermaster supplies.

From his base in Manchester, Minter wrote Lawton in Richmond:

> "I am indirected(?) to apply to Brg Gen McRae Cmdg agent as to aid me, and I am happy to say that he has very promptly agreed to let me have fifteen to twenty thousand suits of clothing, consisting of jacket, pants, shirts, shoes, socks and blanket which Major JB Ferguson will turn over to me as fast as they can be gotten ready for my official receipt for the same. These stores I hope to set off to Havana between the first and middle of November, from which they will go into the Trans Miss Dept with as little delay as possible..."[7]

On November 29 and December 19, 15,492 uniform jackets and 15,500 pairs of trousers were provided by Ferguson. Five thousand five hundred of these trousers and 5,492 jackets were supplied by Hebbert & Co. at a cost of 18 shillings for each uniform. The rest of the suits came from Peter Tait & Co. at a cost of 18 shillings and 6 pennies. The Tait uniforms were shipped from Limerick to Liverpool in two batches of 5,000. Each batch consisted of 68 bales packed in canvas oilcloth and sent on board the steamer *Adelaide* from Liverpool to Havana, Cuba.[8]

CHAPTER 21
HEBBERT & CO.

Charles Hebbert and his business partner established Hebbert & Hume in 1814. The company was described in the Post Office directory as: "Helmet, Army Cap and Accoutrement makers,"[1] based at 30 Prince's Street, Soho, London. The company underwent several name changes before finally becoming Hebbert & Co. in 1852.

Hebbert & Co. also produced military and ceremonial edged weapons. In the National Maritime Museum at Greenwich, London, there is a Hebbert & Co. marked half-basket hilted sword dated 1829 and has the address 8 Pall Mall etched on the blade.

In 1829, the company made uniforms for the newly formed London Police Department. So many officers were sacked for being drunk while on duty that Hebbert finally complained to the authorities about the cost of altering uniforms every time a new officer was hired.

By the 1840s, Hebbert & Co. was one of the largest suppliers to the British Army and was shipping uniforms all over the British Empire. While companies like Peter Tait & Co. embraced the new sewing machine, the Hebbert & Co. business model relied upon independent contract labor called material makers who performed work piecemeal from their homes. This made the company non-competitive in the market.

On March 28, 1856, Hebbert & Co. signed a contract with the Lords of the Admiralty to supply clothing for the British Marines. However, the company was not able to meet the deadline for delivery and applied for an extension. The request was denied. When the uniforms arrived, hundreds of garments were rejected, and a fine of £3,451,12s.4d ($17,255) was levied against the firm. Subsequent deliveries resulted in the rejection of a large number of garments. Uniforms produced in the first four months of 1856 were rejected due to late delivery and poor quality.

The Admiralty commissioned a committee to investigate why Hebbert & Co. had failed to meet the standard of the government sealed pattern. On June 22, 1857, the committee published their reasons for stripping Hebbert & Co. of their government contracts:

> "*Messrs Hebbert & Co. had fallen into submitting articles as dissimilar to the patterns as to cause the committee to believe that no care had been taken in their preparations or measurements. The general inferior quality of Messrs Hebbert & Co.'s goods is too patent to require any opinion of judgment other than which the committee possesses.*"[2]

Fortunately for Hebbert & Co., the Confederacy needed uniforms.

★★★

Ferguson split the Trans-Mississippi Department's order between Peter Tait & Co. and Hebbert & Co. The two companies had worked together in the past to provide uniforms for troops serving in India. The Hebbert contract was composed of the following:

"November 29, 1864

| "5,375 | Military Jackets | 10 shillings |
| "5,350 | Pair Trowsers (sic) | 8 shillings |

"December 19, 1864

| "150 | Pairs Pantaloons | 8 shillings |
| "117 | Jackets | 10 shillings |

| "Grand Total: | 5,492 Jackets | |
| | 5,500 Trowsers"[3] | |

Lots of buttons back marked Hebbert & Co. have been excavated in Texas suggesting that at least some of the jackets made it through the blockade to the Trans-Mississippi Department.

During late 1864 and early 1865, Minter also bought goods directly from Hebbert & Co. One order was for four artillery harnesses at the cost of £30,19s.6d ($150). The following list reveals the goods Minter was looking to purchase and ship.

"24th October 1864
"Hebbert & Co. Pall Mall East, London

"Complete artillery harness including saddles
"Saddles complete including pommel holsters
"Carbine buckets

"Cavalry accoutrements, including black leather 20 round pouches
"Buff leather pouch belt
"Cap pocket, buff

"Infantry accoutrements, including black leather 50 round pouches
"Buff leather pouch belt
"Buff leather waist belt w/ snake hook catch
"Buff leather frog for bayonet
"Buff leather cap pocket to slide on pouch belt
"Expense pouch
"Buff leather rifle sling
"Zinc oil bottles for expense pouch

"Bulk leather – curried and buff
"Black harness hides
"Black harness backs
"Brown bridle middlings
"Black bridle middlings
"Brown harness backs
"Hogskins
"Buff leather middlings for accoutrements"[4]

The End

In the late 1860s, Hebbert & Co. began to diversify and concentrate on the leather supply and accoutrement portion of the army business. Even when the country was not at war, leather belts, scabbards, cartridge boxes and cap pouches still needed periodic replacement. By making this adjustment, Hebbert & Co. survived another forty years.

In 1908, the British Army adopted web equipment, which meant the loss of the leather business. David Hazel, Managing Director of Hebbert & Co., had good connections with the Territorial Associations, which had superseded the Volunteer movement. He approached the Mills Equipment Company and proposed to act as that company's agent to handle sales outside of the London area through the Associations' branches in Glasgow and Leeds. This arrangement lasted until Hazel closed down Hebbert & Co. to start a new firm, Hazel & Co. Ltd. This new company lasted until the outbreak of World War I.

CHAPTER 22
CONFEDERATE JACKET BUTTONS

Buttons used by Peter Tait & Co. for the Confederate contract bore the letters I and A (infantry and artillery). No cavalry buttons were made or supplied. All Tait buttons consisted of the same design, which featured a domed front with the front and back plates held by a U-sectioned band around the outer edge. These buttons were the smaller 3/4" size, as compared to the larger approximately 7/8" sized buttons normally seen on Confederate jackets, although existing Tait buttons vary in size from 3/8" to 3/4".

The floating shank design is the defining feature of all Tait buttons and was found on most of the buttons produced by button makers in Birmingham. Button makers in London and Manchester tended to make buttons with a fixed shank. Tait and Hebbert ordered buttons for their Confederate jackets from Smith & Wright in Birmingham. According to the company's order books and invoices, Peter Tait & Co.'s orders for buttons consisted of the floating shank while Hebbert & Co. opted for the more familiar fixed shank style buttons.[1]

The most popular Tait infantry button was the Roman I button. It had fine horizontal lines in its interior, a domed front, with the front and back plates held by a U-sectioned band around the edge. The back of the button was bordered by a circular dotted line. Stamped within the circle was P. Tait & Co/Limerick or simply P. Tait & Co.

Another type of infantry button found on the jackets was the ornate manuscript I, which had fancy manuscript scrolls filling most of the field. The center of the I was again lined. These buttons were back marked P. Tait/Limerick or P. Tait & Co/Limerick. This button also employed the floating shank and came in two sizes: 3/4" and 7/8".

Tait also used a button with an Old English script I. There were fine horizontal lines within the I. The button had a floating shank and was back marked P. Tait & Co.

Tait only used one type of artillery button. They were exactly the same design as the infantry button except the artillery button had a Roman A on a plain field. The inside of the A was marked with horizontal lines. The button was 3/4" in diameter. The back side was marked P. Tait & Co/Limerick.

Confederate officers' buttons marked *R.T. Tait & Co*/London/Essex St. Strand have also been excavated in the South. These buttons are officer buttons; convex, two-piece type with an eagle on a lined field standing on a pedestal. The size is approximately 1". The buttons are very rare and believed to be the only button supplied by the London subsidiary.

Tait Button Pictures
(From the David Burt Collection)

Hebbert & Co: Confederate Buttons

Hebbert's jackets used only the Roman I and A buttons. They were exactly the same as the buttons Tait supplied, except for the shank. The buttons were back marked Hebbert & Co/London inside a row of dots.

The company also provided buttons for the Confederate Navy, which differed from the infantry and artillery buttons. They were convex, two piece and 3/4" in diameter. The buttons had a design of cross cannons with a fouled anchor device marked CSN. The buttons had a fixed shank and were back marked Hebbert & Co/London.

Hebbert buttons for the CS Navy
(From the David Burt Collection)

Many buttons have been excavated by relic hunters from battlefields and campgrounds. Listed below are the majority of the places where these finds have been made:

* ✯ Lined I Tait & Hebbert buttons have been excavated in fairly large quantities in Texas, including the supply hub of Millican and campsites like Sabinal.
* ✯ Army of Northern Virginia and Army of Tennessee winter camps 1864-65.
* ✯ North and South Carolina battlefields of the 1865 campaigns.
* ✯ Fort Fisher, North Carolina.
* ✯ The Port of Wilmington, North Carolina.
* ✯ The Petersburg siege lines and the Appomattox retreat route in Virginia.
* ✯ Tait buttons have been found at Saylor's Creek battlefield and Appomattox Court House.

CHAPTER 23
CONFEDERATE CONTRACT JACKETS

The design of the Tait jackets made for the Confederacy are the basic 1863 British Army dress tunic minus the skirt. This gave the jacket a tapered short waist and eye catching appearance. The distinctive curved collar on the surviving Tait jackets closely resembles the collars on the 1863 British Army pattern dress tunic. The British Army tunics also had a metal hook and eye fastener. It was not fashionable to have the shirt showing, and the presence of the hook and eye fastener ensured that the top of the shirt collar did not show. Many of the surviving Tait jackets feature this distinctive fastener.

In 1863, Tait supplied 20,000 tunics to the Canadian militia. The uniforms had trimmed epaulettes and an eight-button front. The collar was curved, trimmed in gold and faced in black cloth. The tunic was made of five pieces and a short skirt. The jacket had no back center seam. There was a double line of machine stitching on the button side. Take away the short skirt, and the design is the same as the jackets Tait made for the Confederacy.

A true Tait jacket was machine sewn, had an eight button front, five piece body (no back center seam) and two piece sleeves. The linings were made from Irish linen (a cloth produced from flax) with a vertical pocket in the left breast. The jacket had a double row of white or brown machine stitching on the front button side. The edge of the button side was left raw edged. The buttonholes were handmade from white or brown thread. The bottom edge of the jacket was turned up 1/2". The lining was tucked into the resulting pocket and the raw edge of the cloth was overcast to the lining. This unusual bottom hem was unique to Tait jackets.

The jacket was produced from the same blue-gray broadcloth used to make greatcoats, tunics and trousers for the British Army. The cloth's color was a dark bluish gray, with a strong blue tint. The major differences among Tait jackets were the colored facings and trims.

Goodwin Hem
(From the David Burt Collection)

Presently, at least ten surviving jackets have been identified as Tait jackets. Two of these jackets belonged to Private Garrett Gouge of the 58th North Carolina and Private Hugh Lawson Duncan of the 39th Georgia. Duncan wore his jacket at the surrender of the Army of Tennessee on April 26, 1865, and at his parole in Greensboro, North Carolina on May 1, 1865. Gouge's last service was guarding quartermaster stores in Greensboro. As both his and the Duncan jacket do not appear to have seen any hard service, it is likely that Gouge drew his jacket from Greensboro.

The jackets have blue piping around the top edge of the collar and the edges of the shoulder straps. Both jackets were produced from dark blue-gray broadcloth. Contrary to the vertical slit pockets found in other Tait jackets, the Gouge jacket has an horizontal inner pocket. Both jackets have all eight original Old English script I buttons stamped P. Tait/Limerick.

The Gouge jacket also has the hook and eye fastener at the front of the collar. The British Army size markings at the base of the collar reads 5-10/30-34.

A short piece of lining in the Duncan jacket has been torn away, revealing white stenciled numbered marks on the inside of the uniform. These are most likely pre-cutting marks.

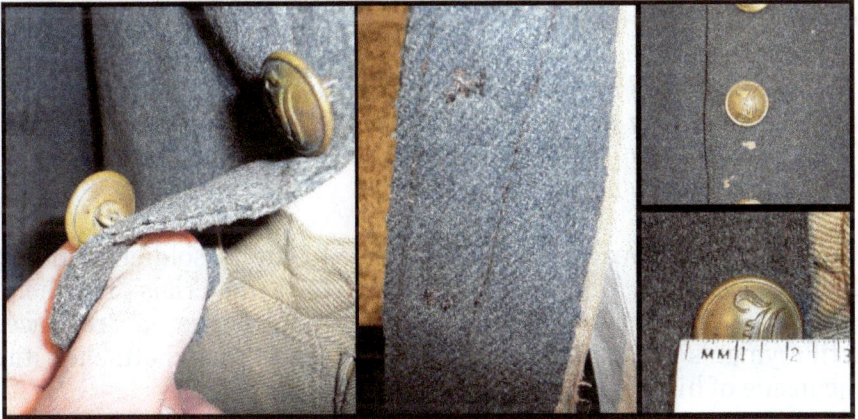

Duncan Jacket
(Courtesy of Dr. Robert Jaffee, M.D)

★★★

One jacket identified as a Tait jacket belonged to Private Benjamin Pendleton of the 2nd Virginia Infantry. Pendleton was wearing the jacket when Robert E. Lee surrendered at Appomattox Courthouse on April 9, 1865. The jacket was made from blue-gray broadcloth and had a red-faced collar also made from broadcloth.

All the buttons on the jacket were replacements: five C.S. staff buttons, one Old English script I button and two Federal Eagle buttons. The buttons were attached to the jacket by square cut nails running through the shank of the button, through the fabric to the inside of the coat. The jacket has a hook and eye fastener. It was machine topstitched in white thread. The buttonholes were hand stitched with white thread.

The Pendleton jacket was assembled by hand, including the buttonholes and the lining. This lining is also of a much lighter shade than in all the other jackets examined.

Another question that has puzzled historians is why an infantryman would be issued an artillery coat, complete with red collar? The answer is found in the Act of Congress of October 8, 1862.

> *"Chapter XXXI*
> *"An Act to encourage the Manufacture of Clothing and*
> *"Shoes for the Army.*
>
> *"Section 4, which stated "That the clothing required to be furnished to the troops of the provisional army under any existing law may be of such kind, as to color and quality, as it may be practicable to obtain, any law as to the contrary notwithstanding."* [1]

The new law made it official practice to supply soldiers with any clothing available, irrespective of color, cut or materials. This helps explains why Pendleton, an infantryman, wore a red-faced artilleryman's jacket. Pendleton wrote a small card and attached it to the inside of his jacket:

> *"This jacket worn by BS Pendleton CO B, 2nd Virginia Infantry, Stonewall Brigade, Army of Northern Virginia, worn at the surrender of General Lee April 9, 1865- BS Pendleton."* [2]

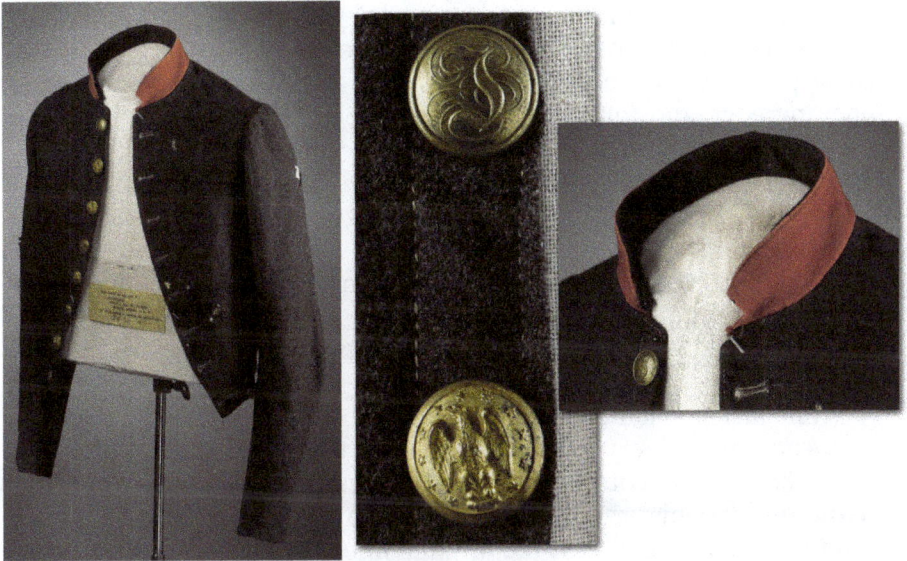

Pendleton Jacket
(Courtesy of the Heritage Auction Galleries)

A letter unearthed in the Fredericksburg and Spotsylvania Military Park was written by Private George L. Slifer. Slifer describes the new jacket he received. "Since we have been down here we bin (sic) supplied with clothing. Drew a nice inglish (sic) suit so you can see I want nothing but our piece and our independence."[3] Slifer belonged to the same regiment as Pendleton.

The hand assembly cast doubts on whether this jacket was made by Peter Tait & Co. It is likely that the jacket was made by Hebbert & Co.

★★★

Another jacket identified as a Tait jacket belonged to Alfred Mercer Goodwin of Sturdivant's Battery, 12th Battalion Virginia, Light Artillery, Army of Northern Virginia. The jacket has solid red flannel facings on the cuffs. These pointed cuffs are on the front of the sleeves and do not extend all the way round. There is evidence that, at one time, the jacket had epaulettes.

The Goodwin jacket has the familiar five piece body, with a 1-3/4" two piece collar with both inner facing and facing material of red wool flannel. The inner facings are 2-1/4" wide at the top and taper to 2" at the bottom hem. There is a five piece lining with seams corresponding to the jacket seams. The lining extends to the front opening edge and forms an inter-lining between the outer layer and inner facing, except at the area where the lining forms pocket flaps. The bottom edge is turned up 3/8" and whip stitched to the lining. It has a double line of machine top stitching in off-white thread 7/8" to 1" wide, nine stitches per inch.

The buttons are Virginia state buttons back marked: Mitchell & Tyler/Richmond VA.

The jacket has two vertical slit pockets in the breast, sized 4-1/2" x 11."

The jacket is hand sewn.

Another mystery is that instead of the usual linen lining, the jacket has a plain cotton and cotton sheeting lining. Les Jenson, who wrote the first history of Tait uniforms, hypothesizes that the cotton lining is a replacement for the original linen lining.

Since the jacket is hand-sewn, doubt must be cast on this jacket being of Tait manufacture.

★ ★ ★

Private William Harrison, a member of Company A, 2nd Maryland Infantry, was issued a Tait jacket on November 13, 1864. Harrison was captured on April 2, 1865, at Petersburg and imprisoned at Point Lookout Prison in Maryland. The condition of the jacket shows considerable wear and tear.

The jacket has a collar face and epaulettes of royal blue broadcloth, which has been described as bright blue. Made of fine quality dark blue gray woolen broadcloth with a linen lining, it is entirely machine sewn – which is evidence that this jacket is of Tait manufacture – and has been topstitched in white thread.

All eight original buttons have been replaced with six Maryland state buttons and two U.S. staff buttons. The buttonholes are hand-stitched in the same white thread used by Peter Tait & Co.

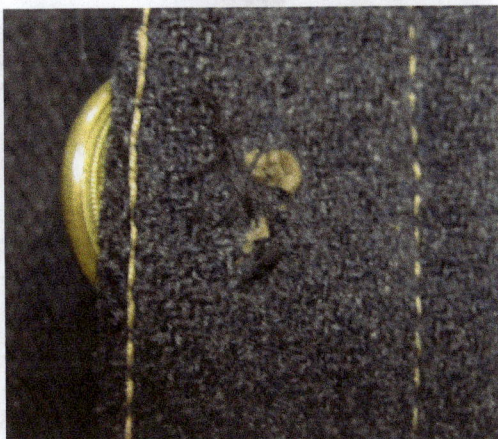

Goodwin Jacket
(From the David Burt Collection)

The British Army size markings read "5-9/38-33." The jacket has a hook and eye fastener.

Harrison Coat
(Courtesy of the Maryland Historical Society)

★ ★ ★

Ordnance Sergeant Marcus De Lafayette Taylor of Company E, 63rd Tennessee Infantry, of A.P. Hill's Corps, also wore a Tait jacket. Taylor surrendered at Appomattox Court House and was paroled on the same day.

The jacket has a royal blue faced broadcloth collar and no epaulettes. The six remaining buttons are original to the jacket. The buttons have the lined Roman I and are stamped P. Tait & Co/Limerick.

The jacket resides in the collection of the Beauvoir Museum in Biloxi, Mississippi.

Taylor Jacket
(Courtesy of Beauvoir, The Jefferson Davis Home and Library)

★ ★ ★

The Museum of the Confederacy has a jacket that has been identified as a Tait jacket. It belonged to Lieutenant Michael G. Glennan. He enlisted in the 2nd North Carolina Artillery and served as an aide-de-camp to Colonel William Lamb, commander of Fort Fisher. Glennan surrendered with the garrison in January 1865. After the surrender, Glennan was hospitalized in Richmond and received his parole from the hospital. There is a possibility he drew the jacket during his stay there.

The jacket features all the characteristics of the Tait jacket, including the five piece bodice, double row of machine topstitching, at nine stitches to the inch. The majority of the jacket is sewn in off white and brown (blue?) thread and is made in the familiar dark blue gray fine kersey.

The collar is 2-1/2" high, with an inner facing of self fabric and a two piece red broadcloth facing applied to the outside. This facing only extends three quarters of the way down the collar.

The sleeves are two pieces with a slightly gathered cap. The sleeves are 7-1/4" wide at the elbow and 5-1/4" wide at the cuff opening. The sleeves feature two piece red wool facings turned under 1/2" and whip stitched to the cuff hem. The cuff hem is 3/4" and sewn to the lining.

The jacket does have a linen lining, which is stamped 6-0/31-36 in black ink with No17 hand written in brown/black ink just underneath the size marks. There are no buttons on the garment.

The jacket has a seven button front instead of the normal eight. The red wool facings on the sleeve cuffs and collar and the fact that that jacket was hand sewn must place doubts on whether the jacket was manufactured by Peter Tait & Co.

Glennan Jacket
(Courtesy of the Museum of the Confederacy)

★ ★ ★

Tait jackets featured British regulation size markings stamped on the original linen lining, just to the right of the collar center seam. For example, the size markings may read 6-0/41-36, which meant the jacket was made for a soldier 6'-0" tall with a chest size of 41" and a waist size of 36". Government contractors produced their British Army tunics and trousers in eight different sizes, ranging from 5'-5" to 6'-0." Chest sizes ranged from 36" to 41" and trouser waist sizes ranged from 31" to a 37" waist. As the 6-0/41-36 size mark in the Glennan jacket indicates, Confederate contract jackets and trousers were marked in British Army sizes, not the traditional American military sizes of No. 1, 2, 3, etc.

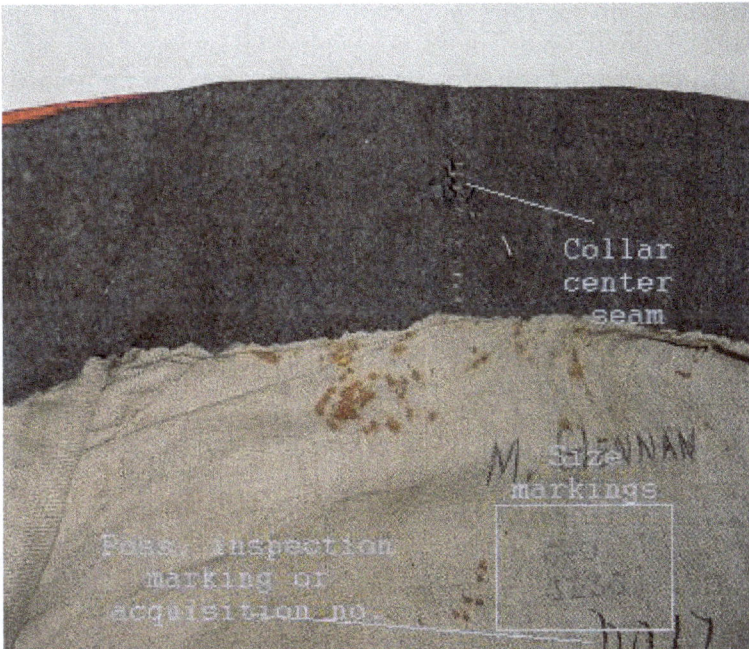

Glennan Coat Size Marks
(Courtesy of Mike McComas)

★ ★ ★

Another jacket in a private collection, with provenance to the Army of Northern Virginia, does display some of the characteristics of the Tait jacket. The exception would be the oddly shaped epaulettes, which are shaped like bowling pins for lack of a better description. The collar is faced in red broadcloth and the epaulettes are made from this same red broadcloth.

209

✯ ✯ ✯

The most unusual Tait jacket known to exist caused a stir at the 2002 Richmond Relic Show. The jacket had been converted into a vest. The owner of the jacket was Francis Goulding of the Jeff Davis Legion, a cavalry unit that served in the Army of Northern Virginia.

Produced of the same fine blue-gray broadcloth as all the Tait jackets and lined in linen, the Goulding jacket/vest has an eight-button front and a royal blue collar. It also has the familiar double lines of white machine stitching and white hand sewn buttonholes. All that remains of the original jacket are the two front and side panels. The rear panel has been replaced with coarse cotton osnaburg.

The two royal blue pocket facings are the same color as the collar. In addition, it appears as though the collar was cut in half to provide the facings for the pockets. The remaining buttons are five cuff-sized New York state buttons.

The buttonholes are original and larger than the cuff sized buttons. The original buttonholes have been sized for 3/4" brass buttons. A considerable amount of effort has gone into this conversion, although the stitching is of poor quality. There is no way of knowing if the alterations were done during or after the war. Since it would haven taken time to alter the jacket in the field, the conversion was probably done post-war for use in civilian life.

✯ ✯ ✯

The final jacket has the strangest story of them all. In the 1950s, a member of the Francis Marion Durham family of Columbia, South Carolina, donated the jacket along with several other stage props and Confederate uniform items to the Colorado Historical Society. Francis "Frank" Marion Durham (1913-1971) was a playwright, actor, director, poet and author whose contribution to South Carolina's 20th century cultural history was significant.

The jacket is in the Tait style and made from blue gray kersey and has all the other attributes, like the double line of machine stitching, lack of center seam, etc., associated with the Tait jacket.

The jacket's curator described the jacket: "It appears that the majority, if not all, of the jacket is hand-stitched. The one area that I am uncertain about is the front placket and the top seam of the collar."4

Goulding Vest
(Courtesy of Chris Daly)

⋆ ⋆ ⋆

The jacket has a linen lining with part of the size stamp mark reading Size No. 2 and a single pocket in the left lapel. It has an eight button front. The only remaining button is a Virginia state button. The collar is red-faced, suggesting that the jacket belonged to an unidentified Virginia artilleryman. The collar has the hook and eye fastener so popular on Tait and Hebbert jackets.

What happened to the jacket after the war or how Durham came to own the jacket is unknown. But since the jacket is primarily hand-sewn, there is considerable doubt as to whether the jacket is of Tait manufacture.

⋆ ⋆ ⋆

Apart from the topstitching on the button side of the jackets, the Goodwin and Glennan jackets are entirely hand sewn as are the Pendleton and Durham jackets. These jackets were made by Hebbert

& Co. Because the Gouge, Duncan, Taylor and Harrison jackets were machine sewn or still have Tait buttons original to the coats, they were manufactured by Tait. Unfortunately, there is not enough information to make a positive identification as to which company manufactured the jacket with the oddly shaped epaulettes and the Goulding jacket/vest.

Durham Jacket
(Courtesy of the Colorado Historical Society,
The Francis Marion Durham Collection)

It is interesting to note that all definite surviving Tait jackets with original Tait buttons have epaulettes, while the jackets lacking Tait buttons do not. Both Tait and Hebbert companies used the smaller 3/4" buttons and, as suppliers of uniforms to the British Army, they would have access to the same uniform pattern. This suggests that the jackets produced by Hebbert and Tait would have been very similar, made from the same patterns and cloth and, as a result, would have been virtually impossible to tell apart, except for the trim details and whether or not the jackets were machine sewn.

✯ ✯ ✯

Another supplier of uniforms and one that has been overlooked is J&J Crombie. Founded by John Crombie in Aberdeen, Scotland, in 1805, the firm produced some of the finest wool and clothing in the country. The Crombie archives from 1861 note that: "Business increased five-fold as Crombie establishes a new export market taking advantage of the lack of mills in the blockaded South to produce the Confederate Army's famous rebel gray cloth."[5] The records state: "The company also produced uniforms for the Confederate Army in the American Civil War."[6]

Any uniforms produced by Crombie would have been based on the design of the British Army coat, also minus the skirt, and the jackets would have been made from the same blue gray cloth as the Tait and Hebbert jackets.

No invoices or shipping records exist stating when Crombie uniforms arrived in the South. But, it cannot be ruled out that maybe one or more of the existing jackets attributed to Hebbert could have been made by Crombie.

CHAPTER 24
WHAT OF TAIT TROUSERS?

In his December 15, 1863 letter to Seddon, James Tait offered to sell to the Confederacy "50,000 Suits of Stout Grey Cloth consisting of jackets and trowsers (sic)."[1] Therefore, when looking for a pair of bona fide Tait trousers, one would have to be looking for trousers the same color and manufacture as the jackets. Royal or sky blue trousers can be safely ruled out.

One of the only known pair of trousers to match the cloth criteria were issued to Lieutenant Glennan.

The trousers' waistband is a 1-1/2" white band, two pieces (left and right) with twilled cotton inner facings. There is a watch pocket set into the top exterior of the band. The seat of the pants feature a twill cotton lining worked into the waistband seam with the loose end turned under 1/8" and sewn with a running stitch. The pockets are 4-1/2" x 1-1/2." The inner facing of flaps were of self-material, with the top edges left raw. There is an orange (originally red?) cotton piping worked into the outside seams from the hem to the pocket flaps. The buttons are 4-hole black horn suspender buttons measuring 5/8" and four, 4-hole black horn fly buttons measuring 1/2".

The most striking detail on these trousers is the use of a 1-1/4" two pronged brass buckle which is stamped L 1859 L. No surviving trouser buckles of CS manufacture have this 1859 stamp. This lends credence to the buckle and trouser being foreign made. The buckle is of superior craftsmanship.

The trousers are entirely hand sewn, which means they could not have been made by the Tait factory.

★ ★ ★

Are there any other surviving trousers that can be attributed to Peter Tait & Co.? One strong candidate surfaced in 1990 at a collectors show in Gettysburg, Pennsylvania. Unfortunately, no record was available of the original owner or anything else that documented their Civil War provenance.[2] Robert McDonald, curator of the Duncan jacket for several years from the late 1980s until the early 1990s, examined the trousers at the show and made an informed opinion based on his considerable expertise and experience. He reported:

> "I saw the trousers in 1990 at the annual Gettysburg ACW Collectors Show. The trousers matched the color and weight of a Tait jacket and the trousers were machine sewn.
>
> "In terms of weave, weight and color, the trousers are an identical match of the wool in the Duncan and Gouge jackets. In my experience, there are apparent differences in the wool between the (few) existing Tait jackets. I believe these variations are more apt to reflect usage/storage/age and light exposure effects rather than initial variations. In all characteristics, the wool in the trousers was extremely similar to that used in the pristine Duncan and Gouge Tait jackets.
>
> "The construction feature I found most interesting was the use of japanned iron rivet-like buttons on the waistband, fly and mule ear pockets. The buttons are certainly the key differentiating characteristic of the trousers. They were anchored rivets, not sewn, with the front of the button being a flat disc atop a short central stud, which was anchored on the reverse by another flat disc, all of them remained very tight."[3]

Japanning was a black glossy coating used to render buttons and other ironware rustproof. The name derives from the method and style that originally came from goods produced in China, India and Japan. In the 19th century, the English West Midlands became important centers for the manufacture of japanned goods.

Rivets themselves were not used in trouser construction until the 1870s, and then rivets were only used on denim work clothes and not on British military uniforms. So, were these rivets a post-war alteration? It's possible, but unlikely, as the rivets appear to be original to the garment.

Since these types of rivet like buttons were not used in any type of uniform in the British Army of the period, it is extremely unlikely that the trousers were manufactured by Tait. McDonald agrees: "I have to clearly emphasize that this pair of trousers is simply a conditional possible candidate for attribution to Tait."[4]

As for the Glennan trousers, the evidence substantiates that they were of foreign manufacture: the matching dark blue gray cloth, the use of the buckle with the 1859 stamp and superior craftsmanship. But like the Glennan jacket, the trousers are completely hand sewn and most likely manufactured by Hebbert & Co.

Unlike the Tait jackets, which came with identifying buttons to prove their provenance, the trousers did not. Therefore, any firm conclusions about the trousers that were exhibited at the 1990 Gettysburg show will remain an unsolved mystery until more specimens can be found and properly evaluated.

Glennan Trousers
(From the David Burt Collection)

CHAPTER 25

PETER TAIT & CO.,

BLOCKADE RUNNERS

It seems to be a common belief that Tait ran his own blockade running fleet to supply uniforms to the Confederacy. This story began in 1954, when J.F. Walsh read the following paper to the Old Limerick Society.

> "Tait directed his attention to the Civil War in America and managed to supply uniforms to the Confederate troops. This was an ambitious undertaking, as distance and transport had to be overcome. However, he solved the problem by purchasing three steamboats, the Evelyn, the Elwy and the Kelpie."[1]

The truth is far different.

The *Elwy* was not built until 1866, a year after the war ended. The *Kelpie* never belonged to or was chartered by Tait. In 1862, after being used as a local ferry in Ireland, the *Kelpie's* owners sent her to the Bahamas, possibly to be used as a blockade runner. On November 28, 1862, *Lloyds List* (the newspaper for the maritime industry) reported that: "The *Kelpie* on her way from Limerick to Nassau had foundered 80 miles east of Nassau, the crew had been saved."[2] On December 8, *Lloyds List* again reported: "The *Kelpie* is reported to have been sunk by (in) a collision 50 miles from Nassau."[3] Neither the *Elwy* nor the *Kelpie* could have participated in Tait's blockade running enterprises.

It was the Alabama contract that caused Tait buy a share in the steamship *Evelyn*. This vessel was the only one that Peter Tait & Co. had an interest in. Tait eventually owned two-thirds of the *Evelyn*.

Tait shipped part of the Alabama order on the *Condor* and the rest on the *Evelyn*. The *Condor* tried to run the blockade and was pursued by a Union gunboat and ran aground off Fort Fisher. Confederate agent, Rose O'Neal Greenhow, a passenger on the

Condor, drowned after the rowboat she was escaping in capsized in the heavy surf. The *Condor*'s cargo was transferred onto the *Annie* and delivered safely to Wilmington.

In Limerick, the *Evelyn* took on board 199 bales of uniforms and 25 cases of army shoes. She arrived in Bermuda in late October, where part of her cargo was off-loaded. She twice sailed from Bermuda, attempting to run the blockade without success and both times had to put back to Nassau. She tried again on December 19, 1864, eventually arriving in Wilmington on December 25 with 170 bales of uniforms.

The 10,000 uniforms provided by Tait & Co. for the Trans-Mississippi Department were shipped on board the steamer *Adelaide,* which left Liverpool on November 14, 1864. These uniforms could not have arrived until the end of 1864 or early in 1865.

★ ★ ★

In 1864, Tait went into business with Alexander Collie & Co. purchasing a portion of the woolen material needed to produce uniforms for the Confederacy. Tait also shipped items on Collie's blockade runners. One recent Tait biographer adamantly contends that Collie cheated Tait by hiding the accounts of profits or sales on the shipments.[4] Whether accurate or not, after the war, Collie presented Tait with a bill for approximately £30,000 ($150,000), his share of their losses, even though Tait had never lost a single shipment to the Union blockade. However, after the losses to their partnership were accounted for on the books, Tait continued to do business with Collie.

CHAPTER 26
AFTER THE WAR

After the war ended, forty-one year old Colonel (compliments of a honorary commission from Robert E. Lee[1]) James Tait married seventeen year old Anabelle Noble, daughter of Benjamin Franklin Noble, senior partner of B.F. Noble & Son Private Bank. Tait had met Anabelle while in Montgomery negotiating the contract for uniforms.

Tait accompanied other mining experts in the investigation of the Birmingham Mineral District. Drawing upon the degree he had obtained from the Bureau of Mines at Oxford University, he predicted that the district would become one of the world's largest industrial centers and wrote several treatises on mining in Alabama. These treatises are part of the state records in Montgomery.

Shortly after completing his investigation, Tait and Anabelle moved to England. In 1876, Tait represented Great Britain at the Paris Exposition. Anabelle became personally acquainted with Empress Eugenie of France and remained friends after the couple's return to Great Britain.

The Taits returned to America and bought a ranch near San Antonio, Texas. James Tait died in 1900 at the age of seventy-six.

★ ★ ★

In December 1866, Peter Tait was elected mayor of Limerick and served three years. During his last year in office, he launched a shipping service between Ostend and Brazil in conjunction with the Belgian government. This concern was known by the magnanimous title: The London, Belgian, Brazil and River Plate Royal Mail Steamship Company. Tait and his old business partner George Cannock were the principal partners in the company.

In 1867, for providing prosperity to Limerick through significant employment, Tait was honored with a sixty-five foot high memorial clock tower in Baker Place, Limerick. He was honored with a knighthood on December 5, 1868, for his contributions to industry and commerce and was known as Sir Peter Tait.

Peter Tait in Mayoral Robes in 1867
(Courtesy of the Limerick City Archives)

In 1868, Tait sought election to the Parliament. He stood for election as a Tory against two candidates of the Liberal party. According to the *Limerick Chronicle*, it was a bitter and disorderly campaign featuring "...factional fighting, bribery and significant damage to property,"[2] along with one death resulting from a pitched fight between rival mobs.

Many citizens were flabbergasted by Peter Tait's association with the Tories – long time adversaries of Irish nationalism – and his opposition to the two Liberal candidates, who had pledged to support Gladstone's popular movement to disestablish the widely resented Anglican Church in Ireland.[3]

Adding to these controversies was the embarrassment of an affair that came to be known as The South Hill Scandal. The scandal involved a nursery maid named Ellen Hinchy and Peter's brother, Robert. In those days, working class girls who became pregnant outside of marriage usually ended up at the local union workhouse. However, this was not the case with Hinchy.

Steps were taken to shield Peter Tait from disgrace by sending mother and child to New York. The Archbishop of New York returned

Hinchy to London after she threatened the life of Peter Tait. Upon her return, she attempted to harm Tait, was accused of the attempted murder of her child and was finally committed to a lunatic asylum. Her child was raised in a Roman Catholic convent. Robert skipped town and was never heard from again.

Tait's political opponents exploited the affair and turned the citizens of Limerick against him. They demonstrated against Tait at the clock tower. Tait resigned as mayor, citing the need "to concentrate on his shipping lines."4

In 1869, the passenger steamship business suffered a downturn and Tait's line collapsed. Tait and Cannock had to liquidate their assets to cover their losses. They sold Cannock, Tait & Co. to Michael J. Clery and James M. Tidmarsh of Clery & Co. Cannock continued in the new business in a management role.

Tait continued on a downward spiral and seemed destined to face further hardship, setbacks and plain bad luck. In October 1872, he became the Conservative candidate for the Orkney and Shetland by-election, which he lost by twenty-six votes. After this final disappointment at the hustings or polls, Tait relocated to London, where little was heard from him for the next fifteen years. Piecing together bits of information, he seems to be chasing military contracts anywhere disagreements were contested by force of arms. His success in securing contracts is not recorded, suggesting there was none to record.

In the meantime, the old Tait & Co. textile factory was sold to a group of British Army officers. It was renamed Auxiliary Forces Uniform & Equipment Co., Ltd. This enterprise continued to operate until 1975. In 2003, most of the building was destroyed in a fire. All that remains today are several outbuildings and part of the original entrance on Edward Street.

In 1886, Tait established the Turkish Cigarette Company Limited in London. Cigarettes were growing in popularity and mild Turkish blends were sought after. However popular cigarettes continued to be, Tait continued his downward cycle. He sold the tobacco business but remained in debt.

In 1890, news reached London that Peter Tait had died on December 15 at the Hotel de France, Batoum, located in the Transcaucasia near the Turkish border. On December 18, 1890, the *London Times* reported that Tait was in the process of setting up another cigarette manufactory in Salonica, Greece. If so, Peter Tait, was a thousand miles from Greece when he died.

Tait bequeathed all his possessions consisting of books, furniture, some linen, as well as the sum of £50 to his wife.[5] Lady Tait passed away at Bournemouth, Dorset in 1906.[6]

★ ★ ★

Sir Peter Tait is best remembered today as a late war supplier to the Confederacy, as well as for the distinctive and stylish jackets of bluish-gray English broadcloth, whether produced in his own manufactory or elsewhere. Surviving Tait jackets bring enormous prices from collectors of the Confederacy. Our research proves that only four of the jackets were made by Peter Tait & Co., while the others were made by either Hebbert or Crombie.

APPENDICES
PETER TAIT & CO.

Appendix O
The Peter Tait Memorial Clock

Tait Clock, Baker Place, Limerick
(Courtesy www.world-stay.com)

This ornamental clock tower, called the Ante-Mortem Memorial by the Dublin builder, was built at Baker Place in Limerick in 1866 as a tribute to Peter Tait, then Mayor of Limerick, for bringing jobs and prosperity to the town. The memorial is a Gothic octagonal tower clock with four faces.

Contributions to build the clock tower came from John Arnott, George Cannock, Alexander Collie and others.

Designed by city surveyor, W.E. Corbett, the tower is inscribed on all four sides.

The west side is inscribed: "Erected by public subscription in appreciation of the enterprise and usefulness of Alderman Peter Tait."

The south side reads: "As an employer of large numbers of the working classes, and of his liberality and benevolence as a citizen."

Tait Clock South Side Inscription
(Courtesy of the Limerick City Archives)

The east side reads: "AD 1866 and completed in the year 1867."

The north side reads: "During Alderman Tait's second mayoralty to which office he was unanimously elected."

The clock was handed over to the Limerick Corporation (Council) on February 21, 1867, and a lavish public banquet was laid out to celebrate the occasion. The Lord Mayor of Dublin, William Lane Joynt, addressed the crowd:

"I assure you I feel the deepest pride and satisfaction in being present on this interesting occasion, which reflects much credit on those whose generosity is manifested in this testimonial, a splendid and enduring proof of their gratitude, and of their kindness towards the Mayor, as it is also proof of his signal services to his fellow citizens. And I am proud that you have departed

from a time honored principle, that you have not waited 'til death laid him low to inscribe on some tombstone a record of his virtues and generosity, but in his lifetime as an encouragement to him to nobler virtues and as a tribute to his greatness."[1]

The clock still stands in its original position. It was refurbished in the early 2000s to run on electricity. It is an important historical monument in the city of Limerick and is a fitting memorial to Sir Peter Tait, the city's most famous entrepreneur and industrialist.

APPENDIX P
LADY ROSE TAIT AND CHILDREN

South Hill House
(Courtesy of the Limerick City Archives)

On June 23, 1853, twenty-four year old Peter Tait married nineteen year old Rose Abraham, daughter of William Abraham and his wife, Elizabeth, both natives of Limerick. The ceremony was conducted by the Reverend William Tarbotten at the Independent Chapel in Limerick.

The couple first lived at 4 Bedford Row, Limerick, but by 1859, moved to the more comfortable surroundings at South Hill House in Rabane, in the County of Limerick. This huge mansion occupied some twenty-six acres of land. While Peter was busy at the Limerick clothing factory, Rose raised nine children at South Hill House and their London address, Mount Pleasant Lodge in Clapton.

Rose stayed by Peter's side during the ups and downs of his career and very little is known about the couple's private life. She became Lady Tait upon her husband's knighthood on December 5, 1868. After his death in 1890, she briefly lived in London and then in

Waverton near Chester, Cheshire. She lived out her remaining days in Bournemouth, Dorset, with daughter Evelyn until her death in 1906 at the age of seventy-two.

The *Limerick Chronicle* reported her death on September 20, 1906:

Death of Lady Tait

"The death took place at Bournemouth within the past few days of Lady Tait, relict of the late Sir Peter Tait, whose memory will forever be revered in Limerick as a philanthropist and an occupant of the civic chair in the far back sixties.

"The deceased Lady Tait had not been in the best of health for some time, having suffered from a malady which necessitated an operation and on which death supervened. Lady Tait was the third daughter of the late Mr. William Abraham, of Fort Prospect of that city, and sister of Mr. William Abraham, Nationalist Member of Parliament for North East division of County. She leaves a family to mourn her demise."[1]

Appendix Q
Army Cloth

All wool is not the same. Different length fibers are woven together to produce varieties of cloths. Heavy woolens are produced from short fiber wool, resulting in a cloth that is bulky, nappy and used for heavier clothing, carpet and blankets. Shoddy cloth is handled much the same as heavy woolens. Worsted wool is produced from long fiber yarn that can be woven tightly. Worsted wool has no nap.

All wool came to the manufactory the same way. The farmer washed and sheared his sheep and sent the wool in bulk to the factory where it was sorted by length of fiber. Long fiber was combed and short fiber was carded. The fibers were spun onto bobbins on a spinning wheel to create yarn and then woven into cloth on a loom. The woven fabric was sent to be fulled – washed and hung up to shrink and dry.

The wool fabric was finished off by shearing the raised nappy areas with enormous four-foot shears, trimming as close to the surface of the cloth as possible. The yarn can be colored at several points in the manufacturing process. Dying the yarn before spinning, results in the most colorfast cloth. The expression "dyed in the wool" comes from fabric dyed at the early point in the process. Wool can also be dyed later, after spinning but before it is fulled.

The cloths used in the Tait contract jackets were dyed after carding, with about 70% blue, 30% gray colorfast dyes. Clean water is essential for proper dyeing. Tait bought his material from J. Ellis & Co., one of the oldest mills in Yorkshire. The company had a large soft water reservoir.

Under examination by Parliament, one of the partners of J. Ellis & Co. stated:

> *"We have an advantage in this neighbourhood... we can dye the finest colors in the world from our water. There is only this to be said, if you economize on the water in cleansing any kind of wool or cloth, you do not*

do the work so well because the more water you have, in reason, the cleaner you get your material... and the better and faster the colors."[1]

Peter Tait and Colin McRae Memorandum

On October 13, 1864, Peter Tait & Co. entered into a direct contract with McRae and Ferguson for the making and delivery of 40,000 uniforms for the Confederate government. The contract read:

"Peter Tait & Co. agree to supply 40,000 (Forty Thousand) Uniforms viz: Jackets and trousers at eighteen Shillings and six pence per suit, net cash, the clothing to be supplied in accordance with the samples deposited with Major JB Ferguson, of assorted sizes, which will be agreed upon by that Officer, and CJ McRae agrees to pay for the clothing in lots of 5,000, five thousand, suits as they are completed. The cloth for these garments to be inspected at Leeds and the made up clothing at Limerick, by Major JB Ferguson or such person as he may appoint for that purpose. The garments are to be packed in waterproof Bales of 150, one hundred and fifty, each, at a charge of 10/6d ten shillings and six pence each bale.

"The carriage of the goods to be paid for at such place, on the production of the carrier's vouchers, as shall be directed by CJ McRae, in addition to the price above named. If shipped at Limerick, no charge for the carriage will be made. The expenses of the inspection at Limerick will be paid for by Peter Tait & Co. who hereby agrees to give every facility to the inspector in the carrying out of his duties.

"This contract is entered into, in the spirit of good fellowship, each party hereto being most anxious that it shall be carried out to the entire satisfaction of all concerned, and is to be completed by the 1st of December 1864.

"Liverpool 13 October 1864

"(Signed) Peter Tait & Co.
"(Signed) CJ McRae"[2]

The contract indicates that Peter Tait & Co. was to supply quantities of blue-gray broadcloth for the Confederate uniforms. On August 3, 1867, The *London Times* newspaper reported that; "the bulk of his (Tait's) cloth is woven in Leeds."[3] Tait opened a new factory on Alexander Street in Leeds in either 1864 or 1865, but very little is known about the factory or what it did.

However, when you consider what the *Times* newspaper wrote in 1867, and the mention of inspecting cloth in Leeds in the Tait and McRae Memorandum, it could be likely that the factory was a weaving mill. If this is the case, then Tait was manufacturing some of his own cloth for the Confederate contracts. From 1865 on, Tait referred to himself as of Limerick, London and Leeds.

Regardless of the source of supply, it can be hypothesized that Tait intended to become the largest supplier of uniforms to the Confederacy. This claim can be substantiated by the October 1864 contract for 40,000 uniforms. Both McRae and Ferguson were happy with both the price and quality of the Tait clothing supplied so far.

Tait Is Sued by the United States Government

In April 1865, Ferguson contracted with Tait for uniforms in exchange for £4,000 ($20,000) worth of cotton. The cotton had been delivered to Tait, but news of the Confederacy's collapse prevented Tait from fulfilling the order. In 1869, the United States government filed a lawsuit against Tait to recover the cotton.

In order for the United States to carry on with the proceedings, they had to show that the Confederacy was a de facto state (exercising power without being legally established) whose property now belonged to the United States. The American Government later dropped the suit.

APPENDIX R
THE ADELAIDE

The *Adelaide*:
Length: 173'-0"
Beam: 26'-0"
Draft: 14'-0"
Registered Tonnage: 300

The ship was owned by Fraser, Trenholm & Co. The *Adelaide* was used to transport supplies out of England to the Caribbean. She was not classified as an actual blockade runner. She carried uniforms to Havana to be transshipped to the Trans-Mississippi Department.

> "*Invoice of Quartermasters' Stores shipped to the Trans-Mississippi Dept by JB Ferguson QM & Major C.S. Army & receipted for by Major JF Minter C.S. Army*"

> "*5,000 Suits Infantry Uniforms at 18/6 each, in 68 bales, made by Tait & Co.*
> "*5,375 "Military Jackets" at 10/ each, and 5,350 pairs "Trowsers"(sic) at 8/ each, in 80 bales by Hebbert & Co.*"[1]

On the same invoice, the *Adelaide* carried another 5,000 suits of infantry uniforms in 68 bales made by Tait & Co., and 150 pairs of pantaloons and 117 jackets in two bales made by Hebbert & Co.[2]

There is no record if the *Adelaide* arrived safely through the blockade, but to judge by the stores that arrived at Brownsville, most of Minter's purchases did get through to the Rio Grande in late 1864 and early 1865. Given that the Trans-Mississippi Department was the last major Confederate force to surrender on May 26, 1865, these uniforms would have had plenty of time to be distributed.[3]

Appendix S
The *Evelyn*

The *Evelyn*:
Length: 270'-0"
Beam: 24'-0"
Draft: 12'-0"
Speed: 15-18 knots
Crew Size: 50

The ship was built by Randolph Elder & Co. in the Govan, Scotland shipyards in 1864, for Alexander Collie & Co. She was described by a Union agent in Bermuda as, "Larger, having three stacks, fore and aft, sidewheels, with a capacity for 1,000 bales of cotton."[1]

The *Evelyn* left Clyde for Limerick on October 25, 1864. She was commanded by Captain H. Talbot Burgoyne. In Limerick, she took onboard 199 bales of the Tait uniforms and 25 cases of boots.[2] She left Limerick for Bermuda on October 27 and arrived on November 16.

The Union Consul in Bermuda reported, "Arrivals from England: *Evelyn*, has cargo. She has been painted white since arrival and will probably leave for Wilmington in a few days."[3] After her arrival in Bermuda, 27 bales of Tait uniforms were offloaded on Collie's instructions. This left a total of 172 bales onboard.

Collie claimed this offloading was done to avoid the risk of the entire shipment being captured by Federal blockaders, but it is believed he put some of his items onboard to sell to the South. The *Evelyn* failed to run the blockade twice. She successfully arrived in Wilmington on Christmas Day, 1864.

On May 15, 1865, an official in the U.S. Naval Department reported: "the following arrivals at Havana all from Galveston: April 21, 1865, *Evelyn*."[4]

News of Lee's surrender reached Cuba, and the *Evelyn* returned to Clyde. Tait eventually bought the ship outright in July 1865 and went on to sell her to the Brazilian Government later that year.[5]

Appendix T
The Condor

The *Condor*:
Length: 27'-0"
Beam: 24'-0"
Draft: 12'-0"
Speed: 18 knots
Crew Size: 50

The *Condor* was also built in 1864 by Randolph, Elder and Company for Alexander Collie & Co. The first shipment of Tait uniforms for the Confederate government and the State of Alabama, totaling some 350 bales, was loaded.

The ship had three raking stacks, fore and aft, turtleback forward, mid ship house, poop deck, two masts, straight stem and painted elusive white – all in all – presenting a striking appearance. She was chased by the USS *Niphon* but arrived safely in Wilmington on October 7, 1864. Her captain tried to avoid the wreck of the *Night Hawk* and ran the ship aground on the Swash Channel Bar at the entrance of the harbor. Her cargo of uniforms and other goods were safely shipped into Wilmington onboard the SS *Annie*.

Until as late as December, lookouts were stationed aboard the ship at low tide. Stranded, with no hope of recovery, Colonel Lamb, Commander of Fort Fisher, used the *Condor* as target practice for his battery. The first shot hit her forward stack, and the second short hit her after stack.[1]

More famous than the ship was one of her passengers. Rose Greenhow was returning home after finishing a speaking tour in Europe. She drowned when the rowboat she was in capsized in the heavy surf. Legend says that Greenhow refused to let go of a leather pouch filled with $2,000 in gold – the royalties from her best selling book on Confederate womanhood.[2] She is buried in Wilmington's Oakdale Cemetery.

Rose O'Neal Greenhow
(Courtesy of the Library Congress)

A newspaper article recounts her funeral:

The Funeral of Mrs. Rose Greenhow

"The death by drowning of Mrs. Rose Greenhow, near Wilmington, North Carolina, last week, has been already noticed. She leaves one child, an interesting little daughter, who is in a convent school in Paris, where her mother left her upon her return to this country.
Hundreds of ladies lined the wharf at Wilmington upon the approach of the steamer bearing Mrs. Greenhow's remains. The Soldiers' Aid Society took charge of the funeral, which took place from the chapel of Hospital No. 4."[3]

A letter to the *Wilmington Sentinel* described the funeral:

"It was a solemn and imposing spectacle. The profusion of wax lights round the corpse, the quality of choice flowers, in crosses, garlands and bouquets, scattered over it, the silent mourners, sable-robed at the

head and foot; the tide of visitors, women and children, with streaming eyes and soldiers, with bent heads and hushed steps, standing by, paying the last tribute of respect to the departed heroine. On the bier, draped with a magnificent Confederate flag, lay the body, so unchanged as to look like a calm sleeper, while above all rose the tall ebony crucifix – emblem of the faith she embraced in happier hours and, which we humbly trust, was her consolation in passing through the dark waters of the river of death. She lay there until two o'clock of Sunday afternoon, when the body was removed to the Catholic Church of St. Thomas. Here the funeral oration was delivered by the Rev. Dr. Corcoran, which was a touching tribute to the heroism and patriotic devotion of the deceased, as well as a solemn warning, on the uncertainty of all human projects and ambition, even though of the most laudable character.

"The coffin, which was as richly decorated as the resources of the town admitted and still covered with the Confederate flag was borne to the Oakdale Cemetery, followed by an immense funeral cortege. A beautiful spot on a grassy slope overshadowed by wavering trees and in sight of a tranquil lake was chosen for her resting place. Rain fell in torrents during the day; but as the coffin was being lowered into the grave, the sun burst forth in the brightest majesty, and a rainbow of the most vivid color spanned the horizon. Let us accept the omen, not only for her, the quiet sleeper, who, after many storms and a tumultuous and checkered life, came to peace and rest at last, but also for our beloved country, over which we trust the rainbow of hope will ere long shine with brightest dyes.

"The pall bearers were Colonel Tansill, chief of staff to General Whiting; Major Vanderhorst, J.M. Seixas, Esq., Dr. de Prossett, Dr. Micks and Dr. Medway. General Whiting and Captain C. B. Poindexter, representing the two services, were prevented from acting as pallbearers, the former by reason of absence, the latter in consequence of illness."[4]

APPENDIX U
THE HOWE AND THOMAS
LOCKSTITCH SEWING MACHINES

Elias Howe was born in Spencer, Massachusetts, on July 9, 1819. After he lost his factory job in the Panic of 1837, he moved to Boston and found work in a machinist's shop. There, he invented the sewing machine.

Howe's machine was the first to employ a needle with an eye. The needle pierced the cloth, creating a loop on the other side. A shuttle on a track slipped a second thread through the loop to create the lockstitch.[1] At 250 stitches a minute, the machine out produced five hand sewers with a reputation for speed.

Howe patented his sewing machine on September 10, 1846, in New Hartford, Connecticut. For the next nine years, he struggled to enlist interest in his machine and to protect his patent from imitators who refused to pay royalties for using his designs.

**Drawing of the Sewing Machine Submitted
to the U.S. Patent Office.**
(Courtesy of the U.S National Archives)

In October 1846, Howe's older brother, Amasa Bemis Howe, traveled to England to seek the financing needed to produce the sewing machine. On December 1, 1846, Amasa was able to sell the patent for £250 to William Thomas of Cheapside, London. Thomas owned a factory that manufactured corsets, umbrellas and valises. Seeing great possibilities for the machine, Thomas employed Elias Howe to produce a machine capable of stitching corsets. Howe failed and left Thomas' employ. But Thomas and his son finally adapted the machine to sew buttonholes, fix hat bands, sew soles on shoes and boots and stitch ships' sails.

Thomas opened factories in London and Birmingham to manufacture his machine for both industrial and home use.

Peter Tait, always ready to embrace any new technology, purchased 150 of the Thomas sewing machines for his factory in Limerick. He testified before the Royal Commission:

> *"I first of all took pains to ascertain the perfection of the sewing machine. Early in 1856, I bought one, fastened it up in my own study and took it all to pieces, and I was satisfied that the machine sewing was better than hand sewing. There is one machine by which if one stitch is cut the work will run, but in Thomas' machine the stitch is perfectly the same on both sides"*[2]

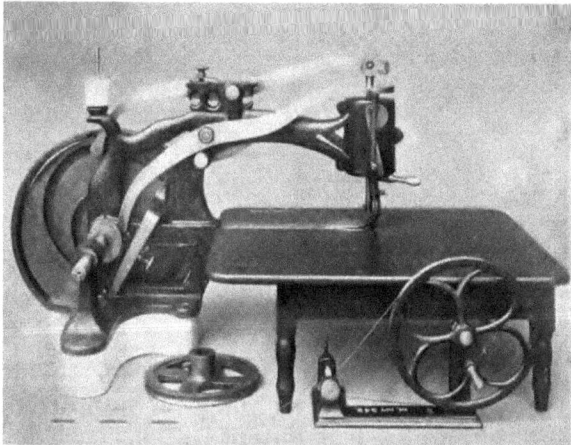

Thomas Lockstitch Sewing Machine, circa 1853
(Courtesy of Grace's Guide, British Industrial History)

The commissioners asked Tait if he could employ machinery to make all the parts of the coat. Tait replied:

"The coats are made by machinery, with the exception of the buttonholes and sewing on of the buttons. An objection has been raised by many other clothiers that sewing machinery is not applicable to many operations, though it may be useful in light descriptions of work. All I have to say is that I have proved the machinery to be perfect. I have gone to a great deal of trouble with it, and I have now brought it to a great state of perfection. I can conduct many branches of business with it. I have a garment here (a sergeant's tunic) which I believe could not be made as well by hand. That badge is put on by steam power, and every stitch, except the buttons and buttonholes, is done by steam power."[3]

Between 1856 and 1858, Peter Tait & Co. supplied 120,000 uniforms for regiments of the line, with total sales of £250,000.

Mr. David Ludlow was the next witness. He had knowledge of the sewing machines used in the Hebbert & Co. factory before the company dispensed with them. The commissioners asked Ludlow about the differences between the two factories.

"Is it not the real difference between Mr. Tait's experience and Messrs Hebbert's that Mr. Tait has thrown his whole soul into the working of this system and carried it out with spirit, and Messrs Hebbert did it under difficulties and never entered very heartily into the adoption of the sewing machine?"[4]

Tait would be proved right in his assumption that the steam powered sewing machines operating in factories was the only way forward. Tait told the commission: "It is a good thing for the country that we have more persevering tradesmen to supply our army now."[5]

APPENDIX V
THE LIMERICK CLOTHING
FACTORY LIMITED

This establishment is justly considered one of the principal sights of this city, and, by the courtesy of the management, is always open for the inspection of visitors. But few, if any, clothing factories in the Kingdom equal it in magnitude, while it stands unique as an Irish industry, both as to the area covered by its buildings as well as in the number of operatives employed.

The premises, originally built for a militia barracks, were secured by the late Sir Peter Tait upwards of thirty years ago, who, with rare foresight, saw that, as far as the Army was concerned, the days of tailor-made garments were doomed, and that, by a sub-division of labour, immense economies were possible. His munificence justly entitled him to the honour he obtained of three times occupying the office of chief magistrate of the city. But few of the original buildings are now in use, as practically, the premises have been rebuilt to render them more suitable to the purposes of the business and will well repay a visit.

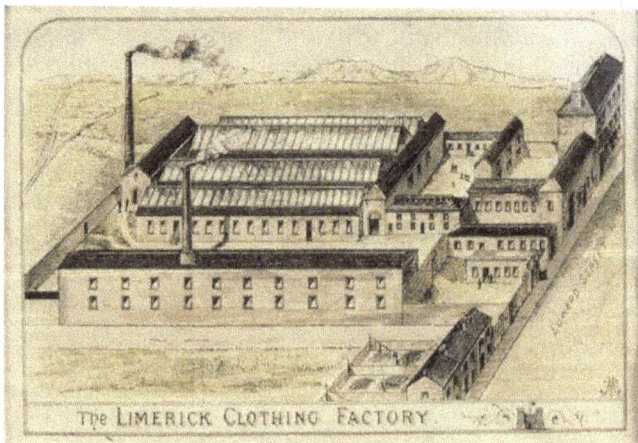

Limerick Clothing Factory

The entrance, beneath a noble archway, introduces us to the court containing the offices, a spacious and convenient building. A second archway leads to the premises of the factory proper. The first factory building we enter is over 300'-0" in length and about 40'-0" wide and is devoted to the reception of the raw material; secondly, rooms for packing (by hydraulic power) and trimming the garments; and, finally, the cutting room.

This room contains seven powerful steam cutting knives, each capable of cutting 672 pairs of trousers in a day. The facility with which a pile of cloth, from twenty to eighty layers in thickness, can be dissected with absolute precision can only be appreciated after seeing the operation. The pattern, or diagram, is marked on the top layer, and the endless band knives follow the lines and divide the block without the slightest difficulty.

From this building we pass into the machine room, of which our illustration (reproduced from a photograph) gives a very inadequate representation, as it has been impossible to give a correct idea of the immense size of the room, which is upwards of 300'-0" in length and nearly 100'-0" in width, roofed in three bays and lighted by a continuous skylight in each bay. The floor is occupied by five long tables, extending the whole length of the building, with wide gangways between. Upwards of 200 sewing machines are at work, requiring a small army of several hundreds of attendants, such as basters, machinists, button-hole makers, finishers and pressers.

Over 700 operatives are engaged in this one room in the production of military garments only. All the machinery is driven by steam power, and, to minimize the risk of accident, the main shafting is all carried in tunnels beneath the floor. Full attention has been given to ventilation, and, notwithstanding the large number of persons in the room, the air is never offensive.

The engines and boilers occupy a building at one end of this human beehive, while rooms at the other are devoted to pressing and repairing shops. The old tailor's goose is conspicuous by its absence, and is replaced by gas-heated irons worked by a system of levers operated by the foot, giving the maximum of pressure with the minimum of exertion.

The engineers shop is replete with all the modern contrivances for repairing machinery. Almost the whole of the work done is paid for by piece, and to such an extent is the division of labour carried, that no fewer than thirty-six different payments are made for the making of certain military garments.

The next room visited is the cap and helmet factory. Here all descriptions of caps and helmets are produced, about fifty persons, male and female, being employed. The wooden blocks upon which the headdresses are shaped are first covered with a cotton lining, and this, having been smeared with a solution of India-rubber, is covered with layers of cork, also treated with the solution, and the whole well amalgamated by repeated hammering with a wooden mallet. The cloth cover, also coated on the inside with India-rubber, is then drawn over the whole, and, after the helmet has been trimmed to shape, bound round the edge with leather, and, its metal ornaments attached, it is ready for issue to the army, police, volunteers or post office, as the case may be. A separate engine and boiler serve this room with heat and power.

Perhaps the most potentially important department is the last to be visited. This is housed in a long two-storied building and is devoted to the manufacture of civilian clothing both for the classes and the masses. The ground floor contains a very large and varied stock, both of ready-made clothing and of cloth, suitable to all purses. The manager is doing all in his power to popularize Irish cloth in Ireland, and so far his efforts have met with marked success, this department being patronized alike by peer and peasant.

The upper floor of this department is a miniature of the large room before described and fitted with steam-power sewing machines and gas pressing irons. Anyone visiting this department must be impressed with the superior sanitary conditions which prevail, as compared with those under which our clothing is usually manufactured in the squalid, unhealthy houses of the operative tailors of the East End of London or the still more unhealthy workroom of the sweater.

The next department to be visited is devoted to providing those of the operatives, who wish to avail themselves of it, with wholesome food at a minimum cost. This room is capable of seating about 200, and is opened for breakfast, dinner and tea. Although the hours worked do not necessitate, under the Factory Acts, any allowance of time for meals, except dinner, the operatives are allowed to break their work for the two other meals if they so desire.

Inside the Factory

This refreshment room was begun by the Most Reverend Dr. O'Dwyer, Bishop of Limerick, and has proved an immense boon to the concern. It is now managed by a committee of the hands, under the control of the manager, and any profit derived is returned to the employees by being handed over to the provident fund.

In order to increase the outlet for the goods manufactured by the company, retail establishments are being acquired in various towns in Ireland, where you can obtain from a suit of ready-made clothing at prices which are really astonishing, or be measured by experienced first-class cutters, and, in either case, receive the benefit in price of the system of sub-division of labour, which has been brought to such a state of perfection.

While, obviously, the commercial element must take precedence in the mind of the head of such an immense concern as this, the social side is not overlooked. A large room is devoted to and fitted up for a theatre and concert room and is capable of seating some 400 audience. Here, at intervals, dramatic and musical performances are given by the employees themselves to crowded audiences. Other rooms are devoted to reading, games and recreation.

Lastly, the management has set itself to solve the problem of provision for old age and infirmity, which question is now exercising the minds of some of our ablest statesmen. All persons throughout the whole establishment voluntarily subscribe, pro rata to their wages, to a benefit fund, and steps are being taken to utilize the capital so obtained in forming and working a co-operative store, the money so used bearing a fixed rate of interest, which, with half the profits of the working of the society, go to form a fund for provision in old age and incapacity. The remaining half, being returned to the purchasers, practically repays the voluntary tax in the wages.

Thus everyone employed has a direct inducement to support the store, as well as having a direct interest in the company, as the surplus capital of the Benefit Fund is invested in the Debentures of the Company. It is calculated that, with practically no diminution of income, the operatives will themselves provide a sufficient competency at an earlier age than that contemplated in the Government Pension Scheme.

A second fund provides a substantial money payment on the death not only of the subscriber but of any near relative of the subscriber. The aim of the manager is to induce the operatives to carry out all schemes of self-improvement themselves, in which he gives them all the assistance in his power.

The present company was established under the Limited Liabilities Acts, in London, and obtained the premises in 1884. The number of persons employed is close upon 1,000 and is by far the largest employment-giving establishment in the city or in the South and West of Ireland.

The registered offices of the company are at Vauxhall Bridge House, Pimlico, London; and the board, under the chairmanship of G.H.M. Ricketts, Esq., C.B., include Messrs. James F.G. Bannatyne, D.L.T. A. Ferguson and O. Wallace, J.P. all from Limerick. The managing director is Mr. E. Taylor, of London, who took over the control of the business in December 1889, and who has since that date increased the number employed by one half and practically reorganized the whole business.

Cannock and Company, Limited

Wholesale and Retail Drapers, Merchant Tailors, Cabinet Makers and Upholsterers, Carpet, Curtain and General Warehousemen of Limerick.

In selecting for this sketch the magnificent temple of commerce which supplies a monumental example of the well-won fruits of Irish commercial enterprise in the City of Limerick, we need offer no apology to our readers for introducing to their notice the colossal undertaking which, under the title of Messrs. Cannock and Company, Limited has achieved a reputation of national importance and a popularity which can only be measured by the immensely successful proportions it has attained during the extended period of its active existence. The history of this notable house dates back to its inception in the early part of the present century, when the business was founded by Messrs. Cumine, Mitchell and Co., Woolen Drapers, this firm being subsequently succeeded in 1850 by Messrs Cannock and White, and later the title was altered to Arnott, Cannock and Co., at the head of which was the present Sir John Arnott, D.L., of Cork, with whom was associated the late Mr. George Cannock.

After some years' prosperous trading, Sir John retired from the firm, and Mr. Cannock was joined by the late Sir Peter Tait, a man of considerable foresight and commercial ability, whose munificent acts as a local public man are remembered with gratitude to this day by the people of Limerick. In 1869, the business was purchased by Mr. J. Clery, J.P., (the present chairman of the company and head of the great house of Clery and Co., Lower Sackville Street, Dublin) and Mr. James Moriarty Tidmarsh, J.P. and was continued by these gentleman with ever-increasing success until 1877, when it was converted into a limited liability company, trading under the title of Cannock and Company, Limited.

Cannock, Tait and Co.

The firm's magnificent premises in George Street are incomparably the most attractive, spacious, and, from every point of view, the most excellent business house in Limerick. It surpasses in architectural grandeur, in extent of area, in convenient arrangement of sections and departments any establishment of the kind in the province of Munster, its noble and graceful clock tower rising to an altitude of 180 feet.

The clock and bells were made and erected in 1888 by the well-known firm of Messrs. Gillett and Johnston (formerly Gillett and Bland) of Croydon, London who also made the large clocks at Sligo Cathedral, Christ Church Cathedral (Dublin) and other important horological works in Ireland. The clock is what is generally known as a Westminster (or Cambridge) chime clock, that is to say, it strikes these chimes every fifteen minutes in addition to striking the hours. The two terms Westminster and Cambridge are synonymous as applied to clocks, and, as a matter of fact, the St. Mary's of Cambridge is the original term for these chimes, although, through greater prominence given to their music since they were introduced into the Westminster Clock Tower, they are more generally referred to as Westminster Quarters in these present times.

In all properly made turret clocks, the weight of the hammers to strike the bells is regulated by the weight of the bells themselves. The reason for this is that, unless the hammer bears a certain proportion to the weight of the bell, it will not bring out its full tone. This proportion is within the discretion of the makers to some extent, but it is generally acknowledged that to bring out the proper tone of a bell the hammer should weigh just that of the bell for the hours, with a somewhat modified figure for the chimes.

Thus we find at Messrs. Cannock's this rule has been well sustained, the weight of the hammers being: 18, 20, 22, 28, and 35 pounds, respectively. These hammers are kept continually off the bells by a powerful steel spring, which is borne down momentarily by the weight of the hammer falling upon it, but regains its normal position directly after the bell is struck and prevents any jarring during vibration.

The clock is now furnished with four dials, a new one having been added by Messrs. Gillett & Co. at Christmas. Each dial is six feet in diameter and glazed with best opal glass, which has the advantage of equally diffusing the illumination behind it, a property not possessed by other species of glass. Each center part is backed with cast-iron ribs to protect the glass from being blown in by high wind. The dials are made of iron, cast in one piece, and painted gilt as shown. The hands of each dial are made out of stout sheet copper, which are stiffened by means of brass run in at the back, which thus prevents them from being blown about and enables them to keep their position on the face. The hands are also balanced inside the tower to preserve their equilibrium.

From the principal entrance in George Street, the visitor passes northwards through the immense range of warerooms, in which are displayed with effective taste the latest fashions in costumes, silks, millinery, mantles, furs, artistic furniture, etc., a total distance of 288 feet; the entire floor area of the establishment occupying the enormous aggregate of 39,360 square feet.

Upwards of 300 sale assistants, clerks and workpeople are employed in the various departments of this mammoth business, in addition to a large number of female hands engaged in the fabrication of the beautiful Limerick lace, a specialty for which this firm is celebrated. Adjoining the wholesale and retail departments, and communicating therewith by means of double iron doors, is the cabinet factory, in which a staff of first-class workmen is employed in the manufacture of high-class furniture, which for artistic design and finish, durability and value, bears favorable comparison with the finest productions of the leading London houses.

To sum up the comprehensive resources in detail of Messrs. Cannock's gigantic establishment would, we fear, be impossible within the brief limits allotted to this sketch, and we must therefore be content to indicate the leading features of their principal departments in addition to those already mentioned. These include woolens, prints, hosiery, gloves, handkerchiefs, haberdashery, ribbons, laces, shoes, silks, cloaks, shawls, flannels, calicoes, perfumery, carpets, hats and caps, bespoke tailoring (a noted specialty), ready-mades, furniture, underclothing, etc.

In conclusion, we may state without exaggeration that Messrs. Cannock and Co. are fully entitled to the designation of Universal Providers in its broadest sense; the babe from its entrance into the world, and in its progress through the various stages of life to its final exit, may here find every requirement for personal comfort or adornment and every article of house furnishing, from which peer or peasant may make suitable selection. Brevity is essential in a review of this character, otherwise we should find emphatic pleasure and satisfaction in dealing more fully with the undertaking and operations of this admirably organized concern, which, under the skilful administration of its able managing director, Philip R. Toppin,

Esq., whose personality is inseparably associated with its signal success, the company is in the unique position of being able to declare at the end of the financial year that the item in their balance sheet headed Trade Creditors is represented by the word nil, or in other words, that they did not at that date owe a single penny for goods supplied.[1]

NOTES

CHAPTER 1. SETTING THE STAGE

[1]William Schouler. *A History of Massachusetts in the Civil War, Volume 1.* (Boston: Dutton & Sons, 1868), 220.

CHAPTER 2. THE BEGINNING

[1]William Albaugh. *A Photographic Supplement to Confederate Swords.* (North Carolina: Broadfoot Publishing, 1993), 233.
Note: This same mistaken background material appears in various other collectors' guides and research anthologies.
[2]"Sale of Property at 71 Jermyn Street, May 1860." *Lincolnshire Archives,* 3/4/2/2.
[3]Ibid.
[4]Janet Elizabeth Kerekes. *Masked Ball at the White Cross Cafe: The Failure of Jewish Assimilation.* (Maryland: University Press of America, 2005), 31.
[5]David Katz. *The Jews in the History of England 1485-1850.* (Oxford: Oxford University Press, 1994), 221.
[6]Ibid., 292

CHAPTER 3. SHOE MAKING IN THE NINETEENTH CENTURY

[1]Men and women employed in the shoe industry worked from their own homes at this time. Richard T. Davenport-Hines. *Capital, Entrepreneurs and Profit.* (London: Frank Cass and Company, 1990), 102.
[2]Van Diemen's Island was a British penal colony. In 1815, the island was renamed Tasmania.
[3]John F. Rees. *The Art and Mystery of a Cordwainer.* (London: Gale, Curtis and Fenner, 1813), 140.
[4]The cordwainers controlled the leather trade back to 1272 and were officially recognized and chartered in 1439. The company motto is *Corietet Arte,* which is Latin for Leather and Art.
[5]The first sewing machine was designed and patented by Elias Howe in the 1840s and later adapted to sewing shoe uppers, which revolutionized shoe making in the United States. The shoes could be sewn inside out, dispensing with the need for a welt or an insole. A machine capable of attaching soles to uppers was patented in 1858 by Lyman Blake, another American inventor. However, soles were still pegged or hand sewn for several years afterwards,

well into the Civil War-era. Although Federal contracts specified sewn soles, pegged bootees were accepted at a reduced rate.

[6]"Mechanisation and Northampton's Shoemakers." (www.bbc.co.uk.).

[7]Ibid.

[8]Ibid.

[9]V.A. Hatley. "Monsters in Campbell Square: The Early History of Two Industrial Premises in Northampton, Northamptonshire Past and Present." *Royal Historical Society, Volume 4.*, 51-59.

[10]Ibid.

[11]Ibid.

[12]The other new shoe manufactory belonged to Moses Philip Manfield. The term labor force as used here means that the firm was setting up in order to innovate and more or less completely adopt the methods of mass production.

[13]Hatley. *Monsters in Campbell Square.*, 51-59.

[14]Turner Brothers, Hyde & Co.'s chief agent in Australia was Mark Coronel. The company placed an advertisement in the *New Zealand Chronicle*, which read in part: "Turner Bros, Hyde and Co having had new and extensive machinery erected for the manufacture of pegged and riveted goods. The introduction of these improvements is enabled to produce above mentioned goods of a better quality and lower prices, in much larger quantities than hitherto." *Western Australia Business Directory, 1864, Nelson Examiner* and *New Zealand Chronicle, 1864.*

CHAPTER 4. WHO WAS CAMPBELL?

[1]John Bigelow. *Retrospections of an Active Life.* (Charlotte: Baker & Taylor, 1909), 449.

Note: President Lincoln appointed Bigelow American Consul in Paris in 1861, Bigelow rose in the ranks from Chargé d 'Affaires to Envoy Extraordinary and Minister Plenipotentiary to the Court of Napoleon III. Working together with Charles Francis Adams, the American Ambassador to the United Kingdom, Bigelow helped to block Confederate attempts to have France and the United Kingdom intervene in the war.

[2]John Bigelow. *Gladstone, Morley and the Confederate Loan of 1863.* (London: DeVinne Press, 1905), 17.

[3]Ibid., See Note 1: "The firm consisted of Samuel and Saul Isaac, though Moses Brothers had a beneficial interest in the cargo."

[4]William Maxwell Evarts. *Arguments and Speeches of William Maxwell Evarts.* (New York: Macmillan Company, 1919), 693. See also: Sessional Papers of the House of Commons, 39. Correspondence respecting the Seizure of the British Vessels *Springbok* and *Peterhof* by United States Cruisers in 1863, Miscellaneous. No. I (1900), C. 34.

CHAPTER 5. THE WEEDON BEC SCANDAL

[1]Michael Jolles. *Samuel Isaac, Saul Isaac and Nathaniel Isaac.* (London: Jolles Publications, 1998), 309.

[2]Cox and Co. began in 1758 when founder Richard Cox was appointed as regimental agent of the Grenadier Guards. The Company is still doing business as Cox & Kings. See: *Cox's & King's: The Evolution of a Military Tradition.*

[3]*The London Examiner*, Saturday July 3, 1858. (British Library, British Newspapers 1800-1900).

[4]*Hansards Parliamentary Papers 1859 II, Volume IX. Hansard Parliamentary Papers, 1859.* Hansard Online Archives (www.hansard.millbanksystems.com).

[5]"Royal Commission of Army Contract." *London Times* (November 1858), *London Times* Archive. (www.thetimes.co.uk/tto/archive).

[6]Ibid.

[7]Ibid.

[8]Ibid., June 30, 1858.

CHAPTER 6. SIC & CO. BOUNCES BACK

[1]Robert Dudley Baxter. *The Volunteer Movement, Its Progress and Wants.* (Cambridge: Macmillan & Co., 1860), 14.

[2]*Fin de siècle* or turn of the century. This term generally refers to the end of the 19th century and beginning of the 20th. The Boer War was fought from 1899 to 1902.

[3]A Clerk of Lieutenancy. *Manual for Rifle Volunteers.* (London: Edward Stanford Publishing, 1859), 14.

[4]"John Bull Guards His Pudding." *Punch* (August 20, 1887). Note: *Punch* was a weekly magazine that ran from 1841 to 2002, specializing in cartoons and satire.

[5]Baxter. *The Volunteer Movement.*, 34.

[6]You could not hold a commission in the British Army if you were Jewish until 1858. Source: Author's discussions with historian, John Hopper.

[7]*Jewish Chronicle.* (April 6, 1860), 4. The *Jewish Chronicle* Online Archives. (www.archive.thejc.com).

[8]A Clerk of Lieutenancy. *Manual for Rifle Volunteers*, 34.

CHAPTER 7. COLONEL JOSIAH GORGAS, CSA

[1]United States War Department. The War of the Rebellion: A Compilation of the Official Records of the Union and Confederate Armies, 128 vols. (Washington, D.C.: Government Printing Office, 1880-1901), Series 4, Volume 1., 220. (Hereafter O.R).
[2]Ibid., 594.

CHAPTER 8. CAPTAIN CALEB HUSE, CSA

[1]"A Rebel Diplomatist-Sketch of Captain Caleb Huse." New York Times, April 10, 1864. New York Times Archive. (www.nytimes.com).
[2]O.R. Series 4, Volume I., 332-33.
[3]S.W. Hoole. Confederate Foreign Agent, The European Diary of Major Edward C. Anderson. (Alabama: Confederate Publishing Co., 1976), 31.

CHAPTER 9. ANDERSON, HUSE & SIC & CO. BEGIN TO SUPPLY THE CONFEDERATE STATES

[1]There were other London commission houses such as Sinclair, Hamilton & Co., that specialized in brokering military supplies. However, SIC & Co. worked exclusively with the Confederate government, whereas others were not as keen to do so, and vice-versa.
[2]O.R. Series 4, Volume 1., 343.
[3]Ibid., 344.
[4]Hoole. Confederate Foreign Agent., 23.
[5]Correspondence Concerning Claims against Great Britain, Volume VI. (Washington, D.C.: Government Printing Office, 1871), 34.
[6]Hoole. Confederate Foreign Agent., 23.
[7]Ibid., 49.
[8]Ibid.
[9]O.R. Series 4, Volume 1., 540.
[10]Hoole. Confederate Foreign Agent., 50.
[11]O.R. Series 4, Volume 1., 1006.
[12]Correspondence Concerning Claims against Great Britain., 79.
[13]Confederate Papers Relating to Citizens or Business Firms 1861-65. National Archives and Records Administration 2133274. (www.archives.gov)
[14]Correspondence Concerning Claims against Great Britain., 103.
[15]O.R. Series 4, Volume 2., 227.,
[16]The Colin McRae Papers, Huse Audit Series. (1861 to 1872), (Columbia: South Carolina Confederate Relic Room and Military Museum Archives, 1861-1872). (hereafter The McRae Papers).

[17]*Correspondence Concerning Claims against Great Britain.*, 79.
[18]Ibid.
Note: George Wythe Randolph (March 10, 1818 – April 3, 1867) was appointed by Jefferson Davis as Secretary of War on March 18, 1862, and took office six days later. He reformed the department by improving procurement and writing a conscription law similar to one he had created for Virginia. He was most known for strengthening the Confederacy's western and southern defenses, but came into conflict with Jefferson Davis over the issue. Due to heath issues, he resigned on November 17, 1862. He died on April 3, 1867, at Edgehill, a family estate in Albemarle County.

CHAPTER 10. RUNNING THE BLOCKADE

[1]O.R. Series 1, Volume 6., 335-356.
Note: An original SIC & Co. invoice in the McRae Papers shows that 6,210 army blankets were to sail on *the Fingal.*
[2]Hoole. *Confederate Foreign Agent.*, 23.
[3]Stephen R. Wise. *Lifeline of the Confederacy: Blockade Running during the Civil War.* (Columbia: University of South Carolina Press, 1988), 53.
[4]*Correspondence Concerning Claims against Great Britain.*, 58.
[5]Ibid., p59.
[6]The McRae Papers.
[7]Ibid.
[8]Ibid.
[9]O.R. Series 1, Volume 1., 202.
[10]O.R. Series 1 Volume 7., 107.
[11]Ibid., 143.
Note: Charles Kuhn Prioleau was a native of Charleston South Carolina. He was manager and partner in the firm Fraser, Trenholm & Co. of 10 Rumford Place, Liverpool and was heavily involved in the purchase of vessels and ancillary goods for the Confederacy.
Note: The *Stephen Hart* was named after Samuel Isaac's father-in-law. Harriet Pinckney was the maiden name of Mrs. Caleb Huse.
[12]Orison Blunt was an appraiser appointed by the U.S. Prize Court. He was a crony of Abraham Lincoln and one of his few supporters in New York City. Blunt had a small Ordnance Department contract to provide the Federal government US-made Enfield P53s rifles. Very few arms were ever produced, and his contract was cancelled when the barrels failed proof. The famous Claud E. Fuller gun collection at Chickamauga/Chattanooga (CHCH) National Battlefield Park has a well-preserved example of a Blunt contract for Enfield rifles on display.
[13]*Correspondence Concerning Claims against Great Britain.*, 93.
[14]Ibid., 111.

¹⁵The McRae Papers.
Note: Middling is a term for leather that is neither good nor bad. It is where the saying "fair to middling" derives.
¹⁶O.R. Series 2, Volume 2., 605.
¹⁷John B. Jones. *A Rebel War Clerk's Diary.* (Urbana: Sagamore Press, 1958), 223.
¹⁸O.R. Series 4, Volume 2., 191.
¹⁹The *Springbok*, 72 U.S. 1 (1866)
Note: The McRae Papers contain SIC & Co. invoices for large quantities of russet brogans.
²⁰Ibid.
²¹Rodman L. Underwood. *Waters of Discord.* (Jefferson: McFarland Publishing Co., 2003), 58.
²²Charles Albert Earp and John B. Tabb. "A Board Confederate Blockade Runners" *America's Civil War.* (Weider History Group, January, 1996).

CHAPTER 11. THE ERLANGER LOAN

¹*Letters Received by the Confederate Secretary of War.* National Archives and Records Administration Record Group 109 G-92-1863. (www.archives.gov).
²O.R., Series 4, Volume 2., 191.
³O.R., Series 2, Volume 2., 642.
⁴Ibid., 641.
⁵Ibid., 642.
⁶*William H. Seward Papers Relating to Foreign Affairs: Diplomatic Correspondence, Volume 1.* (Washington, D.C.: Government Printing Office, 1864), 643.
⁷William Lewis Dayton. *The Civil War: Diaries and Collected Papers.* (Murfreesboro: Middle Tennessee State University Library), MFM98.
⁸Ibid.
Note: How best to explain the attitude expressed by Dayton to Secretary of State Seward? In a word, Anti-Semitic, which was prevalent in the North and limited opportunities for Jews there. The Confederacy was much more accepting of Jews and Jewish culture. See also: Robert M. Rosen. *Southern Jews in the Civil War.* (Columbia: University of South Carolina Press, 2001).
⁹R.A. Brock, Editor. "If We Had the Money." *Southern Historical Society Papers, Volume 35.* (1907), 201-203.
¹⁰*William H. Seward Papers Relating to Foreign Affairs: Diplomatic Correspondence, Volume 1.*
Note: The 14th Amendment of the Constitution reads, "But neither the United States nor any state shall assume or pay any debt or obligation incurred in aid of insurrection or rebellion against the United States, or any

Notes

claim for the loss or emancipation of any slave; but all such debts, obligations and claims shall be held illegal and void."
[11]O.R. Series 4, Volume 2, 889.
[12]Ibid., 888.
[13]Ibid., 645.

CHAPTER 12. FERGUSON AND CRENSHAW

[1]O.R. Series 4, Volume 2., 30-31.
Note: The Confederate War Department was a cabinet-level department responsible for the administration of the affairs of the armies. The War Department was the largest department in the Confederate government.
[2]James Ferguson, Major. "Letter to Alexander Lawton, September 1863." *Compiled Service Records, Fold 3*. National Archives and Records Administration. (www.archives.gov).
[3]Harold Wilson. *Confederate Industry, Manufacturers and Quartermasters in the Civil War.* (Jackson: University Press of Mississippi, 2002,), 157- 160.
[4]The McRae Papers.
[5]O.R. Series 4, Volume 2., 558.
Note: An $2,000,000 appropriation was made during the Confederate Congress' second session (Act 117 approved May 10, 1861) to purchase and construct one or more ironclad vessels-of-war in either France or England. John North was sent to procure them.
[6]Hoole. *Confederate Foreign Agent.*, p63.
[7]Wilson. *Confederate Industry.*, 165.
[8]O.R. Series 4, Volume 2., 826.
[9]Actually it was spilt of 1/8 share to Crenshaw and 1/8 share to Collie.
[10]O.R. Series 4, Volume 2., 601.
[11]The McRae Papers.
[12]Ibid.
[13]O.R. Series 4, Volume 2., 543-547.
[14]Ibid.
[15]Ibid., 893.
[16]Ibid., 564.
[17]*Correspondence Concerning Claims against Great Britain.*, 120.
[18]Ibid., 121.
[19]Ibid.
[20]O.R. Series 4, Volume 2., 886.

CHAPTER 13. THE DISPATCHING OF COLIN MCRAE

[1]O.R. Series 4, Volume 2., 891.

²"Matthias H. Bloodgood's Application for Pardon." *Case Files for Applications from Former Confederates for Presidential Pardon 1865-67. Confederate Amnesty Papers.* National Archives and Records Administration., M1003, Fold 3. (www.archives.gov).
Note: There is no indication of who Bloodgood's friend was, but he was probably a high ranking official within the Confederate government.
³Ibid.
⁴The McRae Papers.
⁵O.R. Series 4, Volume 2., 891,
⁶Ibid., 886.
⁷Ibid.
⁸The McRae Papers.
⁹O.R. Series 4, Volume 2, 889.
¹⁰Ibid. 890.

<h2 style="text-align:center">CHAPTER 14. THE INVESTIGATION</h2>

¹O.R. Series 4, Volume 2., 892.
²The McRae Papers.
³O.R. Series 4, Volume 2., p.885.
⁴Ibid., 894.
⁵Ibid.
⁶O.R. Series 4, Volume 2., 155.
⁷The McRae Papers.
Note: Thomas & Hollams was founded in 1862 by John Hollams (1820-1910). Hollams was articled to a firm of solicitors in Maidstone. In 1840, he came to London and served his articles with the firm of Brown, Marten and Thomas. He was admitted as a solicitor in 1844, and the next year, his firm took him into partnership. By hard work and integrity, he obtained a foremost place in his profession.
⁸O.R. Series 4, Volume 3., 703.
⁹The McRae Papers.
¹⁰"Tait & Co. Invoices." (Austin: Center for American History, Ramsdell Microfilm Collection at the University of Texas).

<h2 style="text-align:center">CHAPTER 15. S. ISAAC CAMPBELL & CO. THE END</h2>

¹*Business Records of Fraser Trenholm.* (Liverpool Maritime Museum), Reel No B/FT 6/19.
²Latin for out of luck.
³Samuel Blatchford. *Reports of Cases in Prize.* U.S. District Court (New York) and U.S. Circuit Court (1866), 427-466.
⁴Ibid. p. 444.

⁵The *Springbok*, 72 U.S. 1 (1866)
Note: In an interesting side note to the case, the written opinion mistakenly refers to Saul as Paul Isaac.
⁶Ellory C. Stowell and H.F. Munro. *International Cases Arbitrations and Incidents Illustrative of International Law, Volume II: War and Neutrality.* (Boston: Houghton-Mifflin, 1916), 388.
⁷In the U.S. Prize Court, the examination of witnesses is held before Court appointed commissioners. The evidence is taken privately.
⁸Stowell and Munro. *International Cases Arbitrations and Incidents.*, 391-395.
⁹Ibid., See also: Wallace: Supreme Court Reports, Volume V, p 21.
¹⁰The *Springbok*, 72 U.S. 1 (1866).
¹¹Court of Bankruptcy, 21 June 1869.
¹²*The Jewish Chronicle,* June 10, 1870. The *Jewish Chronicle* Online Archives. (www.archive.thejc.com).
¹³*The Jewish Chronicle*, July 1, 1870. The *Jewish Chronicle* Online Archives. (www.archive.thejc.com).
¹⁴"Lease from the Duke of Newcastle to Saul Isaac." Nottinghamshire Archives, NCB 3/5/1/3, (June 15, 1875).
¹⁵*The Jewish Chronicle*, October 9, 1903. The *Jewish Chronicle* Online Archives. (www.archive.thejc.com).
¹⁶The archives record several lawsuits that named Saul Isaac as Defendant. The outcome of these High Court cases is not known.

CHAPTER 16. CONCLUSION

¹The library excuse provided by Huse as a way of explaining the acceptance of substantial funds from the Isaacs is puzzling to this day. There is no evidence that Gorgas or Seddon tasked him with the job of procuring a library for the Confederate States and permitting overcharges to do so.
²O.R. Series 4, Volume 11., 889-890.

APPENDIX B. NORTHAMPTON SHOE FACTORY

¹Richard Rove. "Northampton in 1869" (abridged), *Good Words Magazine*, November 1869.

APPENDIX C. BRITISH FIRMS THAT CONDUCTED BUSINESS WITH THE
CONFEDERATE STATES OF AMERICA

[1]Richard I. Lester. *Confederate Finance and Purchasing in Great Britain.*
(Charlottesville: University of Virginia Press, 1975), Appendix IX.

APPENDIX D. HUSE'S BRITISH IMPORTS

[1]O.R. Series, 4, Volume 2., 382-384.

APPENDIX F. ALEXANDER COLLIE & CO.

[1]O.R. Series 4, Volume 3., 525-530.
[2]John E. Waite. *Peter Tait: A Remarkable Story.* (Great Britain: Milnford
Publications, 2005), 49.
[3]Ibid., 60.
[4]Charles Woolley. *Phases of Panic: A Brief Historical Review.* (Great
Britain: Henry Goode & Son Publisher, 1869), 37-38.
[5]Ibid.

APPENDIX H. BOOTS AND SHOES SUPPLIED BY SIC & CO.

[1]Devlin, J. Dacres and Critica Crispiana. *The Boots and Shoes, British and
Foreign of the Great Exhibition.* (Great Britain: Houlston and Stoneman,
1852), 35.
[2]"Le Duc Letter, Lieutenant & Quartermaster, XI Corps." National Archives
and Records Administration., Record Group 92, entry 999.
(www.archives.gov).
[3]John Hopper. *Northamptonshire and Its Part in the American Civil War.*
(Unpublished, 2006).
[4]The McRae Papers.
[5]The *Springbok*, 72 U.S. 1 (1866).
[6]Ibid.
[7]John Crockford. *Reports of the Cases Relating to Maritime Law.* (London:
Court of the Admiralty, 1864-1867), 74-75.
[8]"Invoices of Trans-Mississippi's Quartermaster Department." (Austin:
Center for American History, Ramsdell Microfilm Collection at the
University of Texas, November/December 1865).
[9]O.R. Series, 4, Volume 3., 930.

Notes

Appendix I. Inspection Certificate

[1]Hoole. *Confederate Foreign Agent* ., 49.
[2]Ibid.
[3]O.R. Series 4, Volume 2., 382-84.
[4]O.R. Series, 4, Volume 1., 594.
[5]The McRae Papers.

Appendix J. Major General William Dorsey Pender's Trousers

[1]Craig Schneider, e-mail message to author, May 11, 2012.

Appendix K. Buttons Supplied by SIC & Co.

General information from: Warren K. Tice. *Uniform Buttons of the United States*. (Gettysburg: Thomas Publications, 1997).

[1]The McRae Papers.
[2]The *Springbok*, 72 U.S. 1 (1866).
[3]Ibid.
[4]Ibid.
[5]Ibid.
[6]*A Catalog of Uniforms.* (Museum of the Confederacy Collection, Appomattox, Virginia).
Note: Hughes' frockcoat was made in England and is one of the few officers coats provided by SIC & Co. The coat is complete with SIC & Co. back marked buttons.
[7]The *Springbok*, 72 U.S. 1 (1866)
[8]Francis Henry Upton. *The United States v The Schooner Stephen Hart and Her Cargo.* (New York: Frank Elroy, Steam Boat and Job Printer, 1863).
[9]Ibid., 2.

Appendix L. SIC & Co. Pattern 1853 Cavalry Sword and 1827/45 Cavalry Officer Sword

[1]C.J. Foulkes and E.C. Hopkinson. *Sword, Lance and Bayonet, The Arms of the British Army and Navy.* (London: Arms & Armour Press; Second Edition,1967.), 57.
[2]"British Cavalry in the Crimea." The National Army Museum. (www.nam.ac.uk).
[3]Hoole. *Confederate Foreign Agent.*, 40.
[4]The McRae Papers.

[5]O.R. Series 2, Volume 2., 179.
[6]"Compiled Service Record of Captain O.C Hopkins, 1st Battalion Georgia Cavalry." *Compiled Service Records, Fold 3*. National Archives and Records Administration. (www.archives.gov).
[7]Tim Prince. College Hill Arsenal Civil War Antiques, Nashville. (www.collegehillarsenal.com).
[8]Foulkes and Hopkinson. *Sword, Lance and Bayonet*.
[9]Prince. College Hill Arsenal Civil War Antiques.
[10]The McRae Papers.

APPENDIX M. OTHER WAR MATERIAL SUPPLIED BY SIC & CO.

[1]John Crockford. *Reports of the Cases Relating to Maritime Law*. (London: Court of the Admiralty, 1864-1867).
[2]The McRae Papers.
[3]The *Springbok*, 72 U.S. 1 (1866).
[4]Ibid.
[5]O.R., Series 4, Volume 2., 382-84.
[6]The McRae Papers.
[7]Ibid.
[8]Ibid.
[9]Crockford, *Reports of the Cases Relating to Maritime Law*.
[10]The McRae Papers.
[11]Ibid.
[12]Martin Petrie. *Equipment of Infantry Part IV*. (London: Topographical Staff, Her Majesty's Stationary Office, 1864), 74-75.
[13]*Daily Constitutionalist*. (January-June 1862). (www2.uttyler.edu).
[14]Kevin Dally, email message with author, August 2013.

APPENDIX N. OBITUARIES

[1]*Jewish Chronicle*, November 26, 1886. The *Jewish Chronicle* Online Archives. (www.archive.thejc.com).
[2]*Jewish Chronicle*, October 9, 1903. The *Jewish Chronicle* Online Archives. (www.archive.thejc.com).
[3]*New York Times,* March 13, 1905. *New York Times* Archive, (www.nytimes.com).

CHAPTER 17. A SUIT OF BLUE

[1]Ted Barclay. *Liberty Hall Volunteers: Letters from the Stonewall Brigade*. (Berryville: Rockbridge Publishing, 1991), Letter dated May 26, 1863.

Notes

[2]Augustus Dickert. *History of Kershaw's Brigade.* (Wilmington: Broadfoot Publishing, Wilmington,1990), 26.

[3]Arnold Gates. *The Rough Side of War.* (Othello: Basin Publishing Co., 1987) Quoted in Larry J. Daniel, *Soldiering in the Army of Tennessee.* (Chapel Hill, University of North Carolina,1991), 174.

[4]Thomas Arliskas. *Cadet Gray and Butternut Brown, Notes on Confederate Uniforms.* (Gettysburg: Thomas Publications, 2006), 68.

[5]Ulysses S. Grant. *The Personal Memoirs of Ulysses S. Grant.* (New York: Charles L. Webster & Company, 1885), 320-321.

[6]Frank Rauscher. *Music on the March: A History of the 114th Pennsylvania.* (Harrisburg: State of Pennsylvania Library, 1892), 134.

[7]O.R. Series 4, Volume 2., 134.

[8]The McRae Papers, The S. Isaac & Campbell & Co. sub-series.

[9]Ibid.

[10]"Richard Waller to A.C. Myers." *Records of the Quartermaster Department.* National Archives and Records Administration, Publication M410, Collection of Confederate Records, Record Group 109. (www.archives.gov).

[11]O.R. Series 4, Volume 3, p527.

[12]Ibid.

[13]Ferguson. "Letter from J.B. Ferguson to Alexander Lawton."

[14]"Letters and Payment Vouchers." *Confederate Citizens and Business Files,* National Archives and Records Administration Record Group 109, Fold3. (www.archives.gov).

[15]O.R. Series 4, Volume 3., 674,

[16]Ibid., 955-958.

Note: Colonel Thomas L. Bayne was appointed Chief of the Bureau of Foreign Supplies of the War Department on July 23rd 1863. Backed by new laws that restricted civilian imports, Bayne imported large amounts of military supplies almost immediately.

Note: The 27,648 yards is worked out from Wilson's book, *Confederate Industry,* 344. Wilson calculates that there were 576 yards of woolens per bale.

[17]Arliskas. *Cadet Grey and Butternut Brown,* 62.

CHAPTER 18. PROSPERITY TO THE TRADE OF LIMERICK

[1]Lerwick is the capital and main port of the Shetland Islands, Scotland, and is located more than 100 miles (160 km) off the north coast of mainland Scotland on the east coast of the Shetland Mainland.

[2]"Extract from the minutes of the evidence taken before the Commissioners appointed to inquire in the state of the store and clothing depots at Weedon,

Woolwich and the Tower, December 7, 1858." *Hansard Parliamentary Papers, 1859.*
3"Enterprise in Ireland." *The London Times,* August 3, 1867. *The London Times* Archive. (www.thetimes.co.uk/tto/archive).

CHAPTER 19. A CONTRACT WITH THE CONFEDERATE GOVERNMENT

1Shoddy cloth is made by cutting or tearing apart existing wool fabric from rags or sweepings and re-spinning the fibers into yarn. As this re-working makes the wool fibers shorter, the cloth is decidedly inferior to the original fabric. Shoddy cloth cannot be dyed as effectively and retains the color of the remnants from which it was made, hence the sorting of the remnants by color is especially important.

Shoddy cloth is often confused with a similar Yorkshire wool product known as mungo, which incorporates tailor's clippings with shredded wool.
2Wait. *Peter Tait.,* 62.
3Alexander Lawton. "Letter to Colin McRae, June 28, 1864." *Letters and Telegrams sent by the Confederate Quartermaster General 1861-65,* National Archives and Records Administration, Collection of Confederate Records, Record Group 109. (www.archives.gov).
4O.R. Series, 4, Volume 3, 525-30.

CHAPTER 20. STATE CONTRACTS

1"Good News for Alabama." *Columbus Daily Enquirer,* October 22, 1864. *Columbus Enquirer Archive.* (www.enquirer.galileo.usg.edu).
2Petrie. *Equipment of Infantry,* 74.
3O.R. Series 4, Volume 2., 529.
4O.R. Series 1, Volume 48 (Part 1)., 167.
5Kate Cumming. *A Journal of Hospital Life in the Army of Tennessee from the Battle of Shiloh to the End of the War.* (New Orleans: J.P. Morton & Co., 1866.), 166.
6O.R. Series 1, Volume 22., 946.
7Joseph Minter. "Letter from Joseph Minter to Alexander Lawton, October 14, 1864." *Compiled Service Record of Joseph Minter.* National Archives and Records Administration, 33-35 Fold3. (www.archives.gov).
8"Invoice of Quartermasters Stores shipped to the Trans Mississippi Department." (Austin: Center for American History, Ramsdell Microfilm Collection at the University of Texas), Microfilm, 209B, Part 47.

Notes

CHAPTER 21. HEBBERT & CO.

[1]*London Post Office Directory, 1814.* (London: Critchett and Woods, Fifteenth Edition, 1814).
[2]House of Commons, Committee Reports, June 1857.
[3]"Invoice of Quartermasters Stores shipped to the Trans Mississippi Department.," Microfilm 209B Part 47.
[4]The McRae Papers.

CHAPTER 22. CONFEDERATE JACKET BUTTONS

[1]"Smith & Wright." *Birmingham and London Trade Directories.* (London: Kelly's Trade Directories, 1860) Also, company patterns and invoice files.

CHAPTER 23. CONFEDERATE CONTRACT JACKETS

[1]James M. Matthews, Editor. *Documenting the American South, An Act to Encourage the Manufacture of Clothing and Shoes for the Army.* (Richmond: R.M. Smith, 1862), 69.
[2]Heritage Auctions Ltd, Dallas, Texas. The auction house sold the jacket in 2007.
[3]George Slifer. "Letter to his Family" (Fredericksburg & Spotsylvania National Military Park, Virginia).
[4]Sarah Gilmore at the Colorado Museum Collection, email correspondence with the author.
[5]Crombie Heritage. (www.crombie.co.uk/heritage).
[6]Ibid.

CHAPTER 24. WHAT OF TAIT TROUSERS?

[1]James L. Tait. "Letter to James Seddon, December 15, 1863." National Archives and Records Administration, Collection of Confederate Records, Record Group 109, Volume 18. (www.archives.gov).
[2]Robert McDonald, email message to author, September 2010.
[3]Ibid.
[4]Ibid.

CHAPTER 25. P. TAIT & CO. BLOCKADE RUNNERS

[1]Waite, *Peter Tait.*, 45.
[2]*Lloyd's List* Newspaper, November 28, 1862. http://www.lloydslist.com.

Note: The Registry of Shipping, later renamed Lloyd's Register, printed its first Register of Ships in 1764.
3Ibid., December 8, 1862.
4Waite. *Peter Tait.*, 49.

CHAPTER 26. THE END

1"Obituary of Annabelle Tait." *Tuscaloosa News,* May 13, 1932. (www.tuscaloosanews.com).
2*The London Times,* (January 11, 1869), *The London Times* Archive. (www.thetimes.co.uk/tto/archive).
3The significance of disestablishing the Anglican Church was not an ecumenical issue; rather it meant that the Roman Catholic Church in Ireland no longer had to pay a tithe to the Anglican Church in Ireland.
4Kevin Hannan. "Sir Peter Tait." (1994) *The Old Limerick Journal.*, 26-32.
5*The London Times'* obituary of Peter Tait contains the only mention of the Greek cigarette factory.
6"Obituary of Rose Abraham Tait," (September 20, 1906), *Limerick Chronicles. Limerick Chronicles Archive.* (www.ilovelimerick.ie.com).

APPENDIX O. THE PETER TAIT MEMORIAL CLOCK

1Hannan. "Sir Peter Tait,"26.

APPENDIX P. LADY ROSE TAIT AND CHILDREN

1"Obituary of Rose Abraham Tait." *Limerick Chronicles.*

APPENDIX Q. ARMY CLOTH

1"Parliamentary Papers of Great Britain, Session V." House of Commons Second Report of the Commissioners on Pollution of Rivers (River Lee) Volume XXXIII, Minutes of Evidence. (February 21, 1867), 76.
2"Peter Tait and Colin McRae Memorandum." National Archives and Records Administration, Confederate Citizens File, Group 109, Document 226, M436. (www.archives.gov).
3*The London Times,* August 3, 1867. *The London Times* Archive. (www.thetimes.co.uk/tto/archive.).

Notes

APPENDIX R. THE *ADELAIDE*

[1]Stephen R. Wise. *Lifeline of the Confederacy: Blockade Running during the Civil War.* (Columbia: University of South Carolina Press. 1988), 286.
[2]"Adelaide Bill of Lading." (Austin: Center for American History, Ramsdell Microfilm Collection at the University of Texas.), MF 209B.
[3]James L Nichols. *The Confederate Quartermaster in the Trans Mississippi: The Blockade Runners Texas Connection.* (New York: Percheron Press, 2006), 67-68.

APPENDIX S. THE *EVELYN*

[1]United States War Department. The War of the Rebellion: A Compilation of the Official Records of the Union and Confederate Navies. (Washington, D.C.: Government Printing Office, 1880-1901), Series 1, Volume 10, 601. (Hereafter O.R.N.).
[2]Waite. *Peter Tait.*, 73.
[3]O.R.N., Series 1, Volume 10, 601.
[4]O.R.N., Series 1, Volume 22., 190.
[5]Waite. *Peter Tait.*, 81.

APPENDIX T. THE *CONDOR*

[1]*Dictionary of American Fighting Ships.* (Washington D.C.: Department of the Navy, the Navy Historical Center), 11.
[2]"Rose O'Neal Greenhow Obituary, October 1, 1864." (Durham: Duke University, Rose Greenhow Papers: Alexander Boteler Papers, Special Collections Library).
[3]"Rose O'Neal Greenhow Obituary." *Wilmington Sentinel,* (October 1864) (Durham: Duke University, Rose Greenhow Papers: Alexander Boteler Papers, Special Collections Library).

APPENDIX U. THE HOWE AND THOMAS LOCKSTICH SEWING MACHINES

[1]The lockstitch sewing machine uses two threads, an upper and lower stitch, which locks or entwines together in the hole in the fabric which they pass through. The upper thread runs from a spool kept on a spindle on top of or next to the machine, through a tension mechanism, take-up arm and the needle. The lower thread is wound onto a bobbin, which is inserted into a case in the lower section of the machine below the material.

[2]"Extract of Minutes of Evidence Commission Appointed to Inquire into the State of the Store and Clothing Depots at Weedon, Woolwich and the Tower."
[3]Ibid.
[4]Ibid.
[5]Ibid.

APPENDIX V. THE LIMERICK CLOTHING FACTORY LIMITED

[1]"With a Description of Leading Mercantile Houses and Commercial Enterprises." *Review Past & Present*. (London: Stratten & Stratten Publishing, 1892), 279-288.

INDEX

Index

Index

Emma Hart, 166
England. *See* Great Britain
Erlanger Bonds, 18, 62, 93
Erlanger Loan, 61-62, 78-79, 83, 87, 89, 93, 96, 101
Erlanger, Emile, 60-62
Erlanger, Emile & Co., 60-65, 78, 118
Erlanger, Emile, Jr., 62
Essex, William & Sons, 45, 118
Eugenie, Empress of France, 220
Eugenie, The, 72
Evelyn, The, 218-19, 235, 268

F

Fair, Elisha, 2
Fane, Georgina, 125
Fawcett, Preston & Co., 50, 118
Fergson JB, Jr. Bros & Co. Cloths, Cassimeres and Vestings, 67
Ferguson, John Boswell, Jr., 52, 66-67, 69-71, 73- 75, 80, 83-88, 126, 142, 173-75, 184-85, 190-91, 193, 232-34, 264
Ferguson,D.L.T.A., 246
Field, Parker & Sons, 119
Fingal, The, 43, 47-48, 162, 256
Firmin & Son, 117
First Boer War, 141
First Corps, Army of Northern Virginia, 171-72, 175
First Zulu War, 141
Fleetwood, Patten, 125
Fletcher, Hall & Stone, 118
Forrester, George & Co., 118
Fort Donelson, USS, 54
Fort Fisher (NC), 198, 207, 218, 236
Fort Prospect, Limerick, Ireland, 230
Fort Sumter (SC), 32
Fortnum & Mason, 44
France, 15,-16, 20, 27, 28
    Government Policy of Neutrality, 1
Frankford Arsenal (PA), 34
Frankfurt, Germany, 60-61
Fraser, John & Co., 46
Fraser, Trenholm & Co., 39, 40, 43, 45, 234, 256

Fredericksburg National Military Park, 203
Freed & Co., 119

G

Galveston (TX), 235
Galway Co., 118
Gayle, Amelia, 34
Gayle, John, 34
Gertrude, The, 93-94, 137-38, 162
Gettysburg (PA), 214
Gettysburg ACW Collectors Show (PA), 215-16
Gettysburg, Battle of, 148
Gillet & Co., 247
Gillet and Johnston, 248
Gillett and Bland, 248
Gilliat, J.S., 125
Gilliat, John K. & Co., 118
Giraffe, The, 72
Gladiator, The, 49-50
Gladstone, William Ewart, 125, 221
Glasgow, Scotland, 15, 48, 117, 195
Glennan jacket, 209, 211, 216
Glennan trousers, 216
Glennan, Michael G., 207, 214
Goldsborough, Louis M., 47
Goodman, John D., 2
Goodwin jacket, 203, 211
Goodwin, Alfred Mercer, 203
Gordon Coleman & Co., 117
Gorgas, Joseph, 33
Gorgas, Josiah, 33-35, 39-41, 43-44, 51, 56-57, 67, 74-76, 78-83, 86, 101, 120, 140, 143-44, 162-63, 254, 260
Gouge jacket, 200, 211, 215
Gouge, Garrett, 200
Goulding jacket, 210, 212
Goulding, Frances, 210
Govan, Scotland, 235
Grand Junction Canal, 21
Grant, J.R. & Son, 117
Grant, Ulysses S., 172, 264
Graysbook, 118
Great Britain, 5, 8-10, 15, 20, 24, 27-28, 37, 41-43, 46-49, 60-61, 67,

274

Index

Index

O

Oakdale Cemetery (NC), 236, 238
Old Limerick Society, 218
Old Tug Co., 118
Orkney Islands, Scotland, 222
Orsini Affair, 27-28
Orsini, Felicie, 27
Ostend, Belguim, 220
Ouachita, The, 56
Overed Guerney & Co., 118
Oxford University
    Bureau of Mines, 220
Oxley, J. Stewart & Co., 118

P

Paisley, Scotland, 119
Palmer, William, 96
Panic of 1837, 238
Panic of 1859, 28
Paris Exposition, 220
Paris, France, 60-61, 63, 81, 89
Parker, Charles, 13
Payne, John M., 157
Peacock, George, 125
Pearson, Zachariah C. & Co., 118
Peel, Jonathan, Major General, 26
Pelham, Charles Worsley Anderson,
    Earl of Yarborough, 6
Pender, William Dorsey, 148
Pendleton jacket, 202-03, 212
Pendleton, Benjamin, 201-02
Persia, The, 2
Petersburg (VA), 198, 204
Philadelphia (PA), 34
Pillans jacket, 188
Pillans, Henry, 188
Pimlico, London, 246
Point Lookout Prison (MD), 204
Poole, Dorset, 5
Porter, S.G., 51
Potts and Hunt, 119
Prioleau, Charles Kuhn, 37, 39-40,
    50, 256
Pryse & Redman, 116
Punch, 29

Q

Quilter, Ball & Co., 87, 89, 91, 107,
    109
Quilter, William, 87

R

R&W Aston, 116
Railway Carriage Makers Railway
    Works, 118
Randolph Elder & Co., 234
Randolph, George W., 45, 51, 56, 66,
    235-36
Remarks on the Production of
    Precious Metals and the
    Depreciation of Gold, 15
Reynolds & Son, 44
Richards & Co., 44
Richmond (VA), 34, 41, 56, 61, 66-67,
    70, 79, 128-29, 174, 180, 190, 204,
    207
Richmond Depot (VA), 151, 172, 174-
    75
Richmond Relic Show (VA), 210
Ricketts, G.H.M., 246
Rideout, W.J., 125
Rio Grande River, 234
Robert E. Lee, The, 54
Robinson & Cottum, 118
Robinson & Fleming, 44
Robt Thos Tait & Co., 180
Ross, Alexander, 70
Ross, Alexander & Co., 48, 69, 118
Ross, William, 187-88
Royal Ordnance Depot. See Weedon
    Bec Depot
Ruffin, Frank, 70
Running Pumps (PA), 33
Ryan, Frank T., 171

S

S. Isaac Campbell & Company. See
    SIC & Co.
Salonica, Greece, 223
Sampson, M.B., 125

Samuel Isaac, Army Contractors,
Accoutrement Makers & General
East India Merchants, 18
San Antonio (TX), 220
Sargant, WL & Son, 116
Savannah (GA), 37, 76
Savory & Moore, 44
Saylor's Creek. *See* Sailor's Creek
Battlefield State Park
Schroder, J. Henry & Co., 118
Schuster, Wilson, 125
Scotland, Aberdeen, 116
Scott, W. & Sons, 116
Scott, Winfield, 34
Sea Queen, The, 50
Seddon, James A., 44, 56-57, 70-72,
74-76, 79-80, 83, 86, 89, 91, 126,
162, 180, 184, 214, 260
Selma Arsenal (AL), 78
Sephardim Jews, 7
Sepoy Rebellion, 28
Seward, William H., 60, 62
Seymour, George Edward, 125
Sheffield, South Yorkshire, 119
Shetland Islands, Scotland, 222
Shreveport (LA), 190
SIC & Co., 3-6, 10-11, 13, 15-18, 20,
23, 26-27, 30-32, 40-46, 48-53,
56-57, 61, 62-65, 67-76, 78-79, 82-
83, 86-96, 100-01, 118, 120, 125,
133, 135, 137-38, 140, 142, 144-45,
148-49, 151-55, 159-60, 162-63,
165-66, 172, 254, 261-63
Simons, Isabella, 5, 166
Sinclair, Hamilton & Co., 118
Sir Douglas Fox, 166
Sir William Peel, The, 50
Slifer, George L., 203
Sligo Cathedral, Christ Church,
Dublin, Ireland, 247
Smith & Wright Ltd., 117, 196
Smith, Edmund Kirby, 189-90
Smith, Henry, 97
Smith, Joseph, 116
Smith, Kemp & Wright, 44, 117, 149
Smith, Larkin, 66-67
Sneedsboro (NC), 78
Soho, London, England, 192

Solingen, Germany, 154, 160
Solomans, Henry, 6
Solomon, Catherine, 5
Sonoma, The, 53, 132
South Hampstead, London, 166
South Hill House, 177, 229
South Hill Scandal, the, 221
Southampton, Hampshire, 82
Southwick, The, 49-50, 172
Spain, 129, 130
Spence, James, 125
Spencer (MA), 238
Speyer & Haywood, 53, 93, 133-34,
137
Spotsylvania National Military Park,
203
Springbok, The, 17, 50, 53, 94-95,
132-34, 137-38, 142, 150-52, 162,
252
St. John's New Brunswick, Canada,
137
Stafford, Staffordshire, 9, 10, 14
Stephen Hart, The, 50-51, 53, 93-94,
137-38, 142, 152-53, 161, 163
Stephens, Alexander & Sons, 117
Stonewall Brigade, 171, 202
Straker, S. & Sons, 118
Sturdivant's Battery, 203
Supply, The, 50
Swinburn & Son, 116

T

T.S. & Co., 138
Tait & Logie, 180
Tait Bros & Co., 180
Tait jacket, 176, 188, , 199-201, 203-
04, 206-10, 212-13, 215-16
Tait, Anabelle Noble, 220
Tait, Evelyn, 229
Tait, James Linklater, 180, 183-87,
214, 220
Tait, Margaret, 177
Tait, Peter, 102, 170, 177-78, 180, 184,
186, 207, 218-23, 226-30, 233,
235, 240-42, 247, 265-68
Tait, Peter & Co., 175, 180, 186-88,
191-93, 196-99, 203-04, 206, 208,

Index

211, 214, 218-19, 222-23, 232-34, 236, 241, 267
Tait, R.T. & Co., 118, 180, 197
Tait, Robert, 180, 221-22
Tait, Rose Abraham, 177, 223, 229-30
Tait, Thomas, 177
Tarbotten, William, 229
Taylor jacket, 206, 211
Taylor, Marcus De Lafayette, 206
Texas State History Museum, 164
Thames River, 47
The Honourable Cordwainers Company, 13
The United States Against the Schooner Stephen Hart and Her Cargo, 153
Thiers, M.A., 15
Third Corps, Army of Northern Virgiinia, 206
Thomas & Hollams, 90-91, 96
Thomas, Bryan Morel, 188
Thomas, William, 177, 240
Thompson, James & George, 117
Tidmarsh, James Moriarty, 222, 247
Toppin, Philip R., 249
Trafalgar, Battle of, 28
Trans-Mississippi Department, 92, 142, 152, 175, 189-90, 193, 219, 234
Commissary Bureau, 189
Cotton Bureau, 190
Medical Bureau, 189
Ordnance Bureau, 189
Quartermaster's Bureau, 189
Tranter, William, 116
Trenholm & Co., 38-39
Trenholm Brothers, 38
Turkey, 223
Turkish Cigarette Company Limited, 222
Turner Bros, Hyde & Co., 13, 44, 92, 114, 119, 140, 142
Turner, Richard, 13
Turner, Thomas, 116
Turquand Young, 129

U

United Kingdom. See Great Britain
United States Military Academy, 34, 36, 167
United States Naval Academy, 168
United States Navy, 234
United States of America, 2, 8, 15, 22, 41, 45, 49, 60, 62, 93, 95, 106, 132, 134-36, 162, 167, 172, 218, 233, 251-52, 257, 262
United States Prize Court, 95
United States Supreme Court, 94-5, 137
United States v. *Schooner Stephen Hart,* 137
United States v. *Steamer Gertrude,* 136
United States War Department Ordnance Department, 34
University of Alabama, 36, 168

V

Van Diemen's Island, 8
Vicksburg, the Battle of, 189
Vienna, Austria, 43
Vienna, The, 56

W

W&W Webster, 92
Walker, L.P., 35, 37, 41, 145
Waller, Richard, 172
Walsh, J.F., 218
Ward & Son, 117
Warrington Crescent, Maida Vale, 165
Waverton, Chesire, 229
Weedon Bec Depot, 20-24, 87, 104-07, 109, 111-13
Weedon Bec, Northamptonshire, 20
Welles, Gideon, 47-48
Wentworth, F. & Co., 119
West India Island, 134, 165
Westminister Bank, 129
Westminster Clock Tower, 247
Whitworth, Joseph & Co., 119
Wilkinson, 119

www.ingramcontent.com/pod-product-compliance
Lightning Source LLC
Chambersburg PA
CBHW060316100426
42812CB00003B/796